Schriftenreihe des Energie-Forschungszentrum

Band 61

I0038333

Das EFZN ist ein gemeinsames
wissenschaftliches Zentrum der
Universitäten:

Technische
Universität
Braunschweig

 TU Clausthal

GEORG-AUGUST-UNIVERSITÄT
GÖTTINGEN

Leibniz
Universität
Hannover

CARL
VON
OSSIETZKY
universität OLDENBURG

efzn
Energie-Forschungszentrum
Niedersachsen

Development of a Process for Integrated Development and Evaluation of Energy Scenarios for Lower Saxony

Final report of the research project
NEDS – Nachhaltige Energieversorgung Niedersachsen

The research project 'NEDS – Nachhaltige Energieversorgung Niedersachsen' acknowledges the support of the Lower Saxony Ministry of Science and Culture through the 'Niedersächsisches Vorab' grant programme (grant ZN3043)

April 1, 2015 – July 31, 2019

Energie-Forschungszentrum Niedersachsen
Am Stollen 19A
38640 Goslar
Telefon: +49 5321 3816 8000
Telefax: +49 5321 3816 8009
http://www.efzn.de

Bibliografische Information der Deutschen Nationalbibliothek

Die Deutsche Nationalbibliothek verzeichnet diese Publikation in der Deutschen Nationalbibliografie; detaillierte bibliographische Daten sind im Internet über http://dnb.d-nb.de abrufbar.

1. Aufl. - Göttingen: Cuvillier, 2019

© CUVILLIER VERLAG, Göttingen 2019
   Nonnenstieg 8, 37075 Göttingen
   Telefon: 0551-54724-0
   Telefax: 0551-54724-21
   www.cuvillier.de

1. Auflage, 2019
Gedruckt auf umweltfreundlichem, säurefreiem Papier aus nachhaltiger Forstwirtschaft.

   ISBN 978-3-7369-7118-9
   eISBN 978-3-7369-6118-0

Participating Professors

Prof. Dr.-Ing. Bernd Engel
Technische Universität Braunschweig
Institute for High Voltage Technology and Electrical Power Systems – elenia
Schleinitzstraße 23
38106 Braunschweig

Prof. Dr. Christian Busse
Carl von Ossietzky Universität Oldenburg
Chair of Sustainable Production Management
Uhlhornsweg 49-55
26129 Oldenburg

Prof. Dr. Frank Eggert
Technische Universität Braunschweig
Institute of Psychology, Division of Research Methods and Biopsychology – IPMB,
Spielmannstraße 19
38106 Braunschweig

Prof. Dr. Jutta Geldermann
Universität Duisburg-Essen
Chair of Business Administration and Production Management
Bismarckstraße 90
47057 Duisburg

Prof. Dr.-Ing. habil. Lutz Hofmann
Leibniz Universität Hannover
Institute of Electric Power Systems - Electric Power Engineering Section
Appelstraße 9A
30167 Hannover

Prof. Dr. Michael Hübler
Leibniz Universität Hannover
Institute for Environmental Economics and World Trade
Königsworther Platz 1
30167 Hannover

Prof. Dr. Sebastian Lehnhoff
Carl von Ossietzky Universität Oldenburg
Department of Computing Science – Energy Informatics
Escherweg 2
26121 Oldenburg

Prof. i.R. Dr. Michael Sonnenschein
OFFIS – Institute for Information Technology
Escherweg 2
26121 Oldenburg

Authors

Christoph Blaufuß
Leibniz Universität Hannover
Institute of Electric Power Systems - Electric Power Engineering Section

Prof. Dr. Christian Busse
Carl von Ossietzky Universität Oldenburg
Chair of Sustainable Production Management

Marcel Dumeier
Universität Duisburg-Essen
Chair of Business Administration and Production Management

Prof. Dr. Jutta Geldermann
Universität Duisburg-Essen
Chair of Business Administration and Production Management

Prof. Dr. Michael Hübler
Leibniz Universität Hannover
Institute for Environmental Economics and World Trade

Maren Kleinau
Carl von Ossietzky Universität Oldenburg
Chair of Sustainable Production Management

Henning Krause
Leibniz Universität Hannover
Institute for Environmental Economics and World Trade

Julien Minnemann
Carl von Ossietzky Universität Oldenburg
Chair of Sustainable Production Management

Marvin Nebel-Wenner
OFFIS – Institute for Information Technology

Christian Reinhold
Technische Universität Braunschweig
Institute for High Voltage Technology and Electrical Power Systems – elenia

Jan Sören Schwarz
Carl von Ossietzky Universität Oldenburg
Department of Computing Science – Energy Informatics

Farina Wille
Technische Universität Braunschweig
Institute of Psychology, Division of Research Methods and Biopsychology – IPMB

Tobias Witt
Georg-August-Universität Göttingen
Chair of Production and Logistics

Project Coordination

Julia Seidel
Technische Universität Braunschweig
Institute for High Voltage Technology and Electrical Power Systems – elenia

Table of Contents

1. Introduction

In order to accomplish the climate change goals set by the Paris Agreement and maintain the global temperature increase below 2°C, every aspect of society needs to contribute. One area addressed in the German Sustainability Development Strategy [1], is the goal of affordable and clean energy. The Energy Sources Act (EEG) sets national targets for the power sector and stipulates that at least 80% of electricity production should come from renewable resources in the year 2050 [2]. Although renewable energy concepts are a key element of energy policy today, sole reliance on increasing the share of renewable energies is not sufficient to build a sustainable energy system [3]. Consideration of sustainability criteria, which targets more than the amount of renewable energies, is thus important in evaluating an energy system.

This final report presents the findings of the research project NEDS – Sustainable Energy Supply Lower Saxony. The main research question of the project is *how can a path towards a sustainable energy system for Lower Saxony be found, modeled, and evaluated?* To answer this question, the project was conducted from 2015 to 2019 under the collaboration of eight research institutes. At the outset of this project in 2015, Lower Saxony had an important role model function within Germany as a state with approximately 38% of its electricity coming from renewable energy sources, especially from biogas and onshore wind. Further progress is, however, necessary to reach the goals described in the state's mission statement [4] and to achieve a renewable energy supply by the year 2050, which satisfies sustainability criteria.

There have been other studies to model possible energy system configurations that can achieve the state's targets (especially in [5]), multiple system configurations and paths toward them are possible. This project used methodology developed to analyze and evaluate the configurations. For an overview of the overall project design, see Figure 1. The project team started out by qualitatively describing the energy system for the year 2015 with respect to technical, economic, social, and environmental aspects. The method of scenario planning was applied to develop future energy scenarios for the year 2050. One scenario was selected and quantified resulting in three alternative system configurations. Four reference years were selected to model and simulate the development of the energy system (2020, 2030, 2040, and 2050). In the last step, all of these system configurations in their different pathways were evaluated using multi-criteria analysis.

1

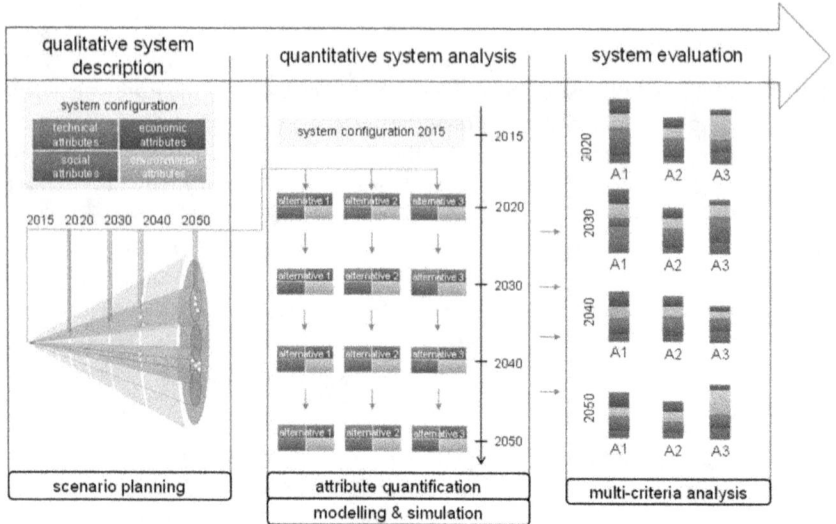

Figure 1: Project overview

In this report, we first present a brief overview of the state of literature (Section 2) before describing the project targets and deriving system boundaries for such a methodology (Section 3). A simple overview of the methodology is given (Section 4), which also encompasses how data exchange was handled in the different phases of the methodology. After that, the methodology is described in more detail. The transition of a formerly centralized electrical energy system with well controllable energy supply toward an energy system based on renewable energies has a significant impact on several areas, e.g., social or economic. The definition of evaluation criteria to assess the sustainability of the defined energy system configurations and transitions paths toward them is presented in Section 0. After that, scenarios are developed in a structured process that also allows the definition of transitions paths (Sections 6 and 8). Selected attributes of these scenarios are empirically embedded, using insights from the diffusion studies, developed for a selection of relevant innovations (Section 7).

With our interdisciplinary project consortium, we aim to analyze technical, social, environmental, and economic parameters as well as their interactions with the help of corresponding models and we focused on coupling these models. Different energy system configurations are modeled using multiple simulations and optimization models described in Sections 9. Results of the individual models are shown in Section 10. Finally, to aggregate the performance scores of the 18 criteria obtained from the simulation and optimization models, a Multi-Criteria Decision Analysis (MCDA) method is used (Section 11).

1. Introduction

To illustrate the application of the method and support the decision process, conceivable transition paths toward a power supply based on renewable energies in Lower Saxony by 2050 were developed and examined for their sustainability and feasibility in several time steps (2020, 2030, 2040, and 2050). With the conclusion of the project, a possible transition path to a sustainable, renewable-energy-based electric power supply system for Lower Saxony was identified for a chosen scenario.

2. State of the Art Regarding Energy Scenarios for Lower Saxony

T. Witt

While many energy scenario studies examine possible future energy systems of Germany [6, 7], there actually are only few relevant energy scenario studies analyzing the transition of Lower Saxony's energy supply system toward higher shares of energy from renewable sources. To decide, which approaches and assumptions can be reused in our project, the existing studies as well as the own approach are categorized using a morphological box.

The method *morphological box* was developed by [8] in the 1960s and can be used to systematically identify all possible configurations of a certain object of interest. In the leftmost column, relevant *parameters* describing a system are collected. For example, in Figure 2, Orientation, Purpose, Type of information, etc. On the right side, the possible variations for these parameters, so-called *characteristics*, are collected. The morphological box is usually used for classification. It can also be regarded as a creativity technique, since new configurations of a system can be found by trying out different combinations of characteristics.

The morphological box for the categorization of energy scenarios has been developed at the beginning of the project [7]. Figures 2 and 3 show the categorization of four related energy scenarios concerning the energy supply system in Lower Saxony as well as the approach in this project (NEDS, marked with *). For example, regarding the scenario orientation, in [9] and [10], *predictive* scenarios are developed, while in this project (*) and [10], *explorative* scenarios are developed, and in [5] and [11], *normative* scenarios are developed. In the following, the other four studies are briefly introduced, before the new features of the approach in NEDS are highlighted.

The study *Szenarien zur Energieversorgung in Niedersachsen im Jahr 2050* (scenarios for the energy supply in Lower Saxony in the year 2050; [5]) was commissioned by the Lower Saxony Ministry for Environment, Energy and Climate Protection, which is a ministry of the federal state of Lower Saxony. Its overall goal was to provide information to develop a guideline for the sustainable development of the energy system, including power, heat, and transport, in Lower Saxony. One important premise of this study is that the whole energy demand of Lower Saxony can be provided with power and heat plants on the actual territory of Lower Saxony, so that land use competition, e.g., between photovoltaics and biogas plants, is considered. This study contains two normative scenarios: The first scenario describes an energy system with 100% renewable energy. The second scenario describes an energy system with a GHG reduction of 80%. Numerical assumptions are presented for 2012 and 2050 only, and most results are only presented for 2050. However, linear extrapolation is used to describe the development of selected key figures in the years 2020, 2030, and 2040.

(1) Scenario properties				
Orientation	Predictive [9, 10]		Explorative (*, [10])	Normative [5, 11]
Purpose	External conditions affecting the consequences of policy actions (*)	Exploration of future conditions or environments [5, 9, 10]	Advocacy of particular courses of action [11]	Representative sample of future states
Type of information	Mainly quantitative [5, 11, 10]		Mainly qualitative [9]	Combined (Story-and-Simulation) (*)
The domain of results/impacts	Technical (*, [5, 9, 11, 10])	Economic (*, [5, 9, 10])	Social (*, [9])	Environ-mental (only GHG) [5, 10]	Environ-mental (*)
Temporal scope of the scenario	Short term			Long term (*, [5, 9, 11, 10])	
Geographical scope of scenario	Local	Regional (*, [5, 9, 11, 10])	National [10]	International	Global [9]
Economic sector	Overall economy (*)	Electricity (*, [5, 9, 11, 10])	Heat [5, 9, 11, 10]	Transport [5, 9, 11, 10]	
(2) Model properties				
Analytical approach / System perspective	Top-down (*)		Bottom-up (*, [10])	n.a. [5, 9, 11]
The geographical scope of the model	Local (*, [10])	Regional (*, [5, 11, 10])	National [10]	Inter-national [10]	Global (*)	n.a. [9]
The temporal resolution of the model	Minutes (*)	Hours (*)	Days [5, 10]	Years (*, [11, 10])	n.a. [9]
Number of models	One [5, 11, 10]		Multiple (*)	
Coupling of models	Soft link (*)	Hard link (*)	No link [5, 11, 10]	n.a. [9]

Figure 2: Morphological box applied to selected energy scenarios

2. State of the Art Regarding Energy Scenarios for Lower Saxony

The study *BUND-Szenario – Energieversorgung in Niedersachsen im Jahr 2050* (BUND scenario – energy supply in Lower Saxony in the year 2050; [11] was commissioned by the Lower Saxony branch of the Bund für Umwelt- und Naturschutz Deutschland (BUND; association for environmental protection and nature conservation Germany). It is based on the scenarios and the energy system model described in [5], but sets different assumptions, for example regarding economic growth, land use of residential areas, resource consumption, and traffic volume. It is also based on the premise that the whole energy demand of Lower Saxony can be provided with power and heat plants on the actual territory of Lower Saxony. This study contains one normative scenario, which describes an energy system with 100% renewable energy. Numerical assumptions and results are described for 2050.

(3) Scientific practice				
Transparency of decision support	Explicit evaluation of scenarios, e.g., with methods from (multi-criteria) decision analysis (*)		Implicit data-driven analysis [5, 9, 11, 10]	
The rationale for assumptions and constraints	Provided, based on literature (*, [5, 10])	Provided, based on own assumptions (*, [11])	Not provided [9]	
Consistency of assumptions and constraints	Demonstrated (*, [10])		Not demonstrated [5, 9, 11]	
Communication of uncertainties	Critical assumptions are marked explicitly (*, [5, 9, 10])		Assumptions are not distinguished [11]	
Ease of model validation	Glass box [5]	Grey box (*, [5, 10])	Black box (*, [9, 11])	
(4) Institutional setting				
Commissioner	Public institution (*, [5, 10])	Private institution [9]	Non-governmental organization [11]	No commissioner
Affiliation of Commissioner	Technical	Economic [9]	Social (*, [10])	Environ-mental [5, 11]	No com-missioner
Involvement of stakeholders	Stakeholders are involved (*, [5])		Stakeholders are either not involved or not mentioned [9, 11, 10]	
*: Approach in NEDS				

Figure 3: Morphological box applied to selected energy scenarios, continued

The study *Energieland Niedersachsen – Struktur, Entwicklung und Innovation in der niedersächsischen Energiewirtschaft* (Energy land Lower Saxony – structure, development, and innovation in the energy sector in Lower Saxony; [9]) was commissioned by the "Institut der Norddeutschen Wirtschaft e.V." (Institute of the North German economy). Notably, this study is not based on any particular energy system model. Therefore, most results are only presented in a qualitative way. Its goal is to provide information on the status quo of the energy sector in Lower Saxony as well as to identify future opportunities for the energy sector in Lower Saxony. This study contains one predictive (reference) scenario, which describes future potentials for the development of Lower Saxony's energy sector, e.g., in terms of the installed capacities of different renewable energy technologies or the development of the number of jobs in the energy sector. Future developments are described for different years up to 2038.

The study *Energie und Klima als Optimierungsproblem am Beispiel Niedersachsen* (Energy and climate as optimization problem applied to Lower Saxony; [10]) was commissioned by the Federal Ministry for Education, Science, Research, and Technology. Notably, this study was published in 1996 and is therefore not only the oldest of the analyzed studies but was also written well before the energy transition became a prominent topic on the political agenda in Germany. It is nonetheless included in the analysis because it is one of the few model-based studies concerning Lower Saxony's energy supply system. The goal of this study is to analyze options, which allow avoiding the use of nuclear energy and reducing CO_2-emissions in Lower Saxony, with an optimization model. This study contains three scenarios and additional sensitivity analyses, which differ in the assumed prices for energy carriers, sociodemographic and economic parameters, and assumptions concerning the use of nuclear energy. Quantitative assumptions are described for 1992, while quantitative results are presented for 2005 and 2020.

In this project, we extend upon existing studies. In particular, the integration of methods from Multi-Criteria Decision Analysis (MCDA) to evaluate transition paths and the integration of energy system models with different foci requires a new methodology for energy scenario development and evaluation. This methodology will be outlined broadly in Section 4. The multi-criteria evaluation of transition paths is described in more detail in Section 11. Another feature of this new methodology is that the consistency of assumptions and constraints is addressed with an information model, which can be used to support the data exchange in the different phases of the developed methodology. For example, when different energy system models with different temporal resolutions and system perspectives need to exchange data, the information model can be used to model the dependencies and construct a shared database for these energy system models (see Section 4.2). Furthermore, the internal consistency of the explorative scenarios is addressed in the scenario planning method (see Section 6). Finally, not only greenhouse gas emissions are considered as environmental impacts, but also other factors such as agricultural land use, by means of a Life-Cycle assessment (see Sections 9.9 and 10.5).

3. Project Framework

C. Reinhold, T. Witt

Because energy systems are very complex systems (as stated in [12]), the major question, how such a system can be transformed so that it is more sustainable in the future, can be broken down into many small research questions that need to be tackled by different disciplines [13]. Increasingly, interdisciplinary approaches, where models and methods from different disciplines complement each other, are used.

In this project, models and methods from energy technology, psychology, business administration, economics, and computer science are brought together to evaluate how sustainable energy supply can be achieved for Lower Saxony up to the year 2050. The project team comprises partners from universities located in Lower Saxony.

In Section 3.1, the project targets are described in more detail. From these project targets, the system boundaries are derived in Section 3.2.

3.1 Project Targets

C. Reinhold

The transformation of the energy supply system in Germany and especially in Lower Saxony toward a more sustainable system requires the investigation of relevant subsystems and properties that describe this system. In this field, essential instruments for decision support and scientific policy advice are the qualitative analysis of future scenarios via scenario planning, quantitative analysis of these scenarios' consequences via energy system analysis, and a subsequent multi-criteria evaluation, which helps to integrate objective model results and stakeholders' values for a holistic system evaluation. In this project, we develop a new general methodology for the development and evaluation of energy scenarios that aim to integrate scenario planning, energy system analysis, and multi-criteria analysis.

The objective of developing this methodology is, therefore, to provide an instrument for scientific policy advice, which helps to shed light on today's decision problems by providing quantitative data and allowing for sensitivity analyses, which can inform corresponding debates in the democratic system and make them more objective and transparent. The methodology does however not claim to generate binding recommendations. The energy transition affects many stakeholders because the energy system has many dependencies to economy, environment, policy, technology, and citizens. Therefore, the methodology has an additional goal to integrate this interdisciplinarity. In the first step, we develop future energy scenarios for the year 2050 as well as transition pathways marking the years 2020, 2030 and 2040. In the second step, we aim to evaluate the final system state in 2050 as well as the transition states based on a sustainability concept using a multi-criteria decision analysis approach.

In addition to the overall goal, subject-specific key questions are formulated at micro- and macro-level. For the micro-level, influencing factors on the diffusion of innovations as well as the possible necessity of behavioral adaptation for the implementation of a sustainable power supply system are examined. This requires a detailed empirical description of the behavior patterns in German households. The integration of innovations and new technologies in the residential and commercial sectors requires the investigation of the electrical behavior of buildings for each grid node in a coordinated grid system. Based on this, technical requirements are defined, which are necessary for the efficient use of smart grid technologies.

The macro-level, on the other hand, analyzes the energy system transformation and its quantitative effects on the sub-sectors of grid technology, energy economy, and national economy. To this end, the topology of the supply system and the characteristics of the generation and consumption structure are addressed. Cost-optimized expansion strategies and economic planning of the plants are essential for the transformation of the power supply system. The focus at the macroeconomic level is the evaluation of the influence of climate policy, trade policy and their interconnections on the energy sector and the economy of Lower Saxony.

Both levels are linked with each other. Statements and results from the micro-level are transferred to models of the macro-level with the use of scaling. For example, the load assumptions of a low-voltage grid are scaled and transferred to the grid extension planning for Lower Saxony's entire power grid.

Answering the key questions for future scenarios requires subject-specific simulation models to simulate the individual parts of the energy supply system. The basis for efficient simulation studies is the performance optimization of the respective simulation models. The subsequent process step is the aggregation and connecting of the simulation results with the sustainability criteria.

3.2 System Boundary

T. Witt

Because the project targets are not only to develop a methodology for development and evaluation of energy scenarios but also to apply this methodology to planning the transition of Lower Saxony's power generation system toward higher shares of energy from renewable sources, *two* system boundaries need to be distinguished:

First, the system boundary of the developed methodology and, second, the case-specific system boundary of its application to Lower Saxony.

The *system boundary of the developed methodology* is related to the so-called life cycle of energy scenarios. According to Grunwald [14], this life cycle can be distinguished into three phases. (1) Construction of energy scenarios, which may include quantitative energy system analysis, qualitative assumptions, or a mix of both; (2) evaluation of energy scenarios, which is the evaluation of an energy scenario's substance[1]; (3) impact of energy scenarios in energy policy and the energy sector, which comprises energy scenarios' consequences for decisions, the formation of opinions or structuring of public debates. In accordance with the project targets, the developed methodology for development and evaluation of energy scenarios covers only the first two phases of an energy scenario's life cycle. The generated results of the methodology can subsequently be used in the last phase of the life cycle to inform public debates. For example, as part of the project, the assumptions and results have been discussed in three public symposia with stakeholders employed in Lower Saxony's ministries, different regional associations, and the energy sector.

The *case-specific system boundary of the methodology's application* to Lower Saxony can again be divided into two parts: system boundaries of the future scenarios (see Section 6) and system boundaries of the different (energy system) models (see Section 9).

The *future scenarios* cover the power supply system of Lower Saxony, e.g., in terms of the energy mix and energy demand, and its environment, e.g., in terms of prices for primary energy carriers and political developments in the European Union. The heat and transport sectors are not covered, but assumptions for the development and growth of the overall economy are made. The temporal scope of the scenarios is 2015 through 2050, and the geographical scope of the scenarios is (mostly) Lower Saxony. For example, the energy demand of the adjoining federal states of Bremen and Hamburg is also considered, following the principle of solidarity described in [5].

Different *energy system models* cover different parts of the power generation system and have different geographical scopes. For example, the energy demand and generation of (a limited number of) households are modeled in a smart grid model. Thus, only a small section of the energy system is modeled, and this model's results

---

[1] According to [14], the substance or "quality", i.e., the assumptions and methods of energy scenarios should be evaluated, which should help to select suitable energy scenarios for a decision problem. Note that a slightly different concept of evaluation is used in this project: Different alternatives are evaluated under different scenarios, which also leads to more transparent decision support. We also aim to increase the transparency of an energy scenario's construction phase with the help of an information model (see Section 4.2).

need to be aggregated before they can be used as input parameters for connected models with greater geographical scope. For detailed descriptions of each model's system boundary and the interfaces between the models, refer to the individual models' descriptions (see Section 9). Due to the varying availability of data, the temporal baseline of the models also differs. For example, while most models use input data from 2015 as their base year, the macroeconomic model uses data from 2011 as its base year, since data that is more recent was not available.

It should be noted that some of the used models and methods, e.g., the multi-criteria evaluation or empirical studies of the diffusion of innovation, require the identification of relevant stakeholders. With the object of investigation being Lower Saxony's energy supply system, different stakeholders can be investigated. However, a decision maker (or group thereof), who could implement options for Lower Saxony's future energy supply system, cannot be clearly identified, because there are simply too many actors that have different authorities over necessary resources (including energy supply companies, transmission system operators, non-energy companies, the general public, government institutions, and non-government organizations). In such a setting, we suggest that the different stakeholders can be analyzed via stakeholder analysis (see, e.g., [15] or Section 7), or that different perspectives are investigated via sensitivity analysis (see Section 5.1 for the weighting of decision-relevant criteria).

4. Development of a Methodology for the Sustainability Evaluation of Energy Scenarios

T. Witt, J. S. Schwarz

The major objective of energy policy in the European Union is to ensure a competitive, sustainable, and secure energy supply [16]. These targets are often cited in energy scenario studies for the EU and its member states. However, most energy scenario studies do not differentiate between scenarios (describing external uncertainties) and alternatives (describing options for the energy supply), and therefore do not explicitly evaluate the degree to which different alternatives actually satisfy these (multiple, usually conflicting) targets [7, 17]. Therefore, the recommendations and conclusions in these studies cannot be derived in a transparent way. To increase the transparency of the decision support, it would be useful to explicitly evaluate options in terms of these targets and thereby support today's decisions in energy policy or the energy sector.

To achieve such an evaluation, the methodology is based on the story-and-simulation (SAS) approach [18], in which the first step for developing scenarios is to develop qualitative storylines for each scenario (via scenario planning, see, e.g., [19]), which are quantified with assumptions and system models (via energy system analysis, see, e.g., [20]) afterward. Our methodology extends this SAS approach with multi-criteria analysis (see, e.g., [21]) in such a way that it leads to a number of alternatives, which can be evaluated in terms of competitiveness, sustainability, and energy security. Figure 4 illustrates the basic idea of a sustainability evaluation process.

The methodology, therefore, combines three individual processes: scenario planning, (energy) system analysis, and multi-criteria analysis. It does however not cover the last phase of the life cycle of energy scenarios (see Section 3.2), due to the nature of multi-criteria decision problems, where different stakeholders may have different opinions and priorities so that a compromise needs to be found.

Future Scenarios		Sustainability Evaluation
Possible future energy system configurations		Sustainability evaluation of decision alternatives
Qualitative future scenario description	----?----	Sustainability order of decision alternatives

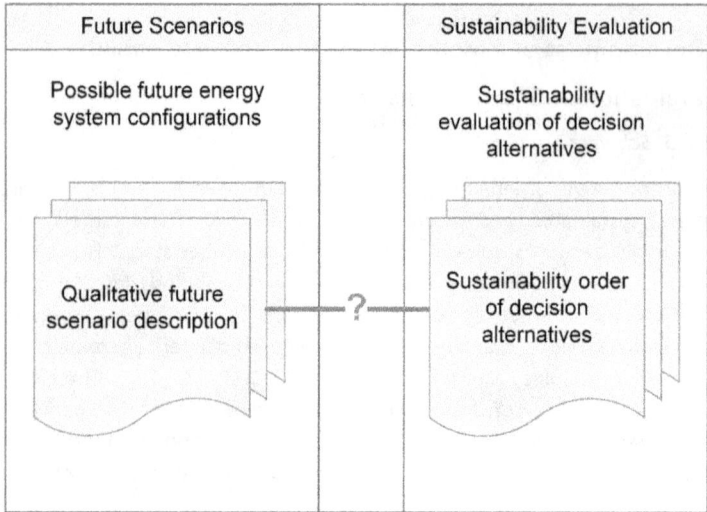

Figure 4: Methodological challenge: from qualitative future scenarios to an order of quantitatively evaluated alternatives

4.1 Process for Integrated Development and Evaluation of Energy Scenarios (PDES)

T. Witt

Two similar versions of the developed methodology have been described in [22, 17] (for a simplified version, see Figure 5; for a detailed version, see Appendix: Figure 101).[1] According to these descriptions, the methodology is divided into four subsequent phases: (1) Preparatory Steps; (2) From Story to Simulation; (3) Modeling and Simulation; (4) Sustainability Evaluation.

In the *Preparatory Steps* phase, a problem is identified and defined. The first result of this phase is a problem adequate definition of sustainability and its operationalization with concrete sustainability evaluation criteria (SEC, see Section 0). For example, in the application of the methodology to planning the transition of Lower Saxony's power generation system, not only social, economic, and environmental, but also technical criteria are used. The second result of this phase is the description of qualitative future scenarios addressing the earlier defined problem (see Section 6). For planning the transition of a complex system such as the energy supply system, different perspectives need to be integrated [13]. Therefore,

---

[1] Another version of the methodology, which uses the individual process of scenario planning, energy system analysis, and multi-criteria analysis as a theoretical basis, is described in [61].

assuming an interdisciplinary perspective can be useful when defining future scenarios.

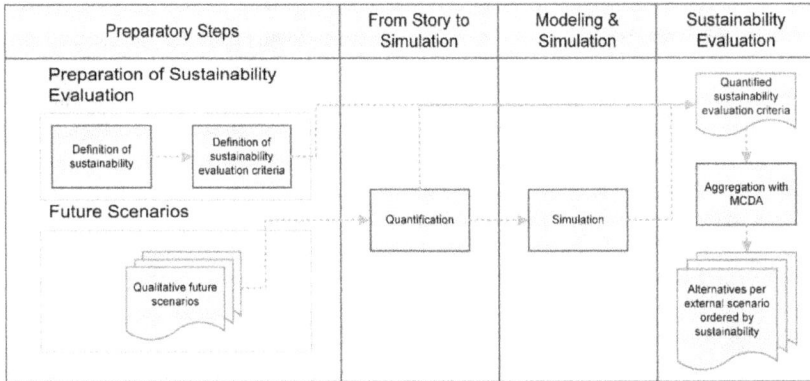

Preparatory Steps	From Story to Simulation	Modeling & Simulation	Sustainability Evaluation
Preparation of Sustainability Evaluation			Quantified sustainability evaluation criteria
Definition of sustainability --▷ Definition of sustainability evaluation criteria			▼
			Aggregation with MCDA
Future Scenarios	▷ Quantification ---▷	Simulation ---▷	▼
Qualitative future scenarios --▷			Alternatives per external scenario ordered by sustainability

Figure 5: Overview of the Process for Integrated Development and Evaluation of Energy Scenarios (PDES)

In the *From Story to Simulation* phase, the qualitative future scenarios are transformed into quantitative assumptions. Assumptions that – according to the case-specific system boundaries – represent external uncertainties, e.g., prices for crude oil, are distinguished from assumptions that represent alternatives, e.g., alternative configurations of the energy system. Together, both types of assumptions form a so-called "evaluation object", i.e., a particular alternative in a particular scenario. The assumptions, e.g., the installed capacity of rooftop photovoltaic power plants, are usually based on related literature, but can also be refined with empirical analyses, as we did in our diffusion studies (see Section 7). The result of this phase is, therefore, the specification of input parameters for quantitative energy system models, which represent the earlier defined future scenarios (see Section 8).

In the *Modeling and Simulation* phase, one or multiple energy system models can be used to quantify the performance of different alternatives in different scenarios (see Sections 9 and 10). In this project, multiple models are used, so that their interfaces need to be defined as well.

In the *Sustainability Evaluation* phase, the alternatives are explicitly evaluated in terms of the earlier defined sustainability evaluation criteria under different scenarios. With sensitivity analyses, the consequences of different input parameters, e.g., the criteria weights, on the evaluation result can be investigated. Based on explicit evaluation of alternatives, policy recommendations can be inferred in a transparent way.

4.2   Information Model

J. S. Schwarz

In the PDES, many different types of data are described as the output of different steps of the process, e.g., quantified future scenarios, results of the simulation, or SECs. Thus, an information model has been developed, which assists the handling of data flows and dependencies during the integrated development of future scenarios, simulation, and sustainability evaluation within the PDES based on methods from computer science. It is described in detail in [22, 17] and summarized in the following.

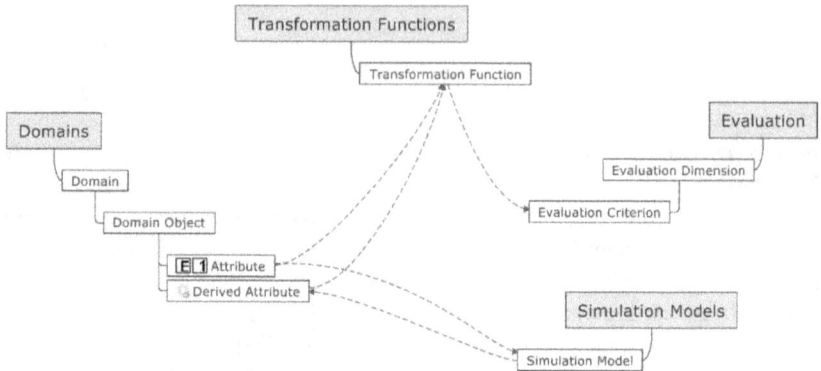

Figure 6: Structure of the information model [17]

The structure of the information model is shown in Figure 6, which uses the object-orientation paradigm from computer science. The boxes with rounded edges represent the different types of objects and the rectangular boxes represent the instances of the objects. Arrows model data flows.

On the left-hand side of Figure 6, the domains of interest within the future scenarios and simulation are modeled. The term domain refers to a specific area of expertise in the interdisciplinary context. Each domain is described by domain objects representing objects or concepts of the real world. These domain objects are characterized by attributes, which can have different types as shown in Figure 7. The attributes are categorized based on the following two characterizations: Firstly, if attributes can be influenced, they are *endogenous* attributes, which constitute the decision alternatives. Secondly, the attributes that cannot be influenced describe the external uncertainties and are subdivided into *scenario-specific framework conditions* and *general framework conditions*. The scenario-specific framework conditions vary for all the different future scenarios, while the general framework conditions apply for all future scenarios.

Additional information can be annotated to the attributes: E.g., the key factor describing the attribute in the scenario planning process (see Section 6.1) or the different types of attributes: general framework condition (G), scenario-specific framework condition (S), and endogenous attribute (E). The attributes can be annotated with their unit and have to be instantiated with values for simulation. The general framework conditions have to be initialized with one value for all external scenarios and decision alternatives. The scenario-specific framework conditions have to be initialized with one value for each external scenario, and the endogenous attributes for each external scenario and decision alternative. These values can be based on literature research for future scenarios (non-derived attributes) or on simulation (derived attributes). For the derived attributes, which are annotated with cogwheel icons, also simulation models can be added in the information model to represent the data flows of their outputs.

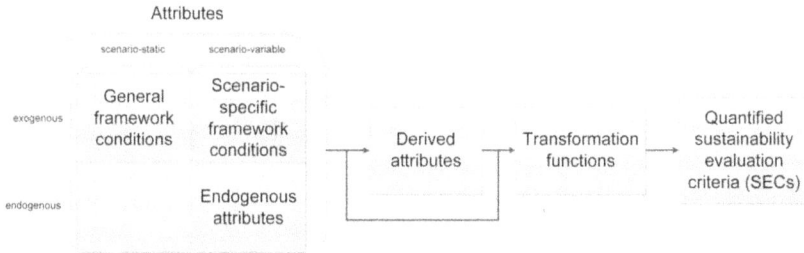

Figure 7: Overview of attribute classification in different types and the data flows of their values [22]

On the right-hand side of Figure 6, the evaluation goal is described – for the NEDS project, this is sustainability. This main goal of the evaluation is subdivided into several sustainability dimensions (technical, economic, environmental, and social as described in Section 0), which are again subdivided into concrete SEC. Quantified attributes are mapped onto the SECs via transformation functions. These functions describe simple data flows or aggregations of the input values.

In Figure 8, an example of the usage of the information model from [17] is shown. The domain *user* contains the domain object *lifestyle* with an attribute *utilization frequency of electrical equipment* representing user behavior. This attribute is scenario-specific (annotated with an "S") and depends on a specific factor defined in the future scenarios (annotated with a "6"). It represents conceivable future development of the user behavior and is input for a smart home model. This model consists of various models of consumer devices encapsulated in a smart home. Based on the user behavior, the building model offers *flexibility*, which is used by the smart grid model to schedule the Building Energy Management Systems' (BEMS) electricity consumption or production. Based on the schedule, the *demand for electricity in households* is the output of the building model, which is input for the economic market model and the energy market model. Both market models also have the *energy mix* as input. The energy mix consists of the amount of energy generated by the different primary energy sources. As output, the two market models provide the *carbon dioxide emissions* of the power plants and the overall economy. The *carbon dioxide emissions* are input for the transformation function for the SEC *GHG emissions*.

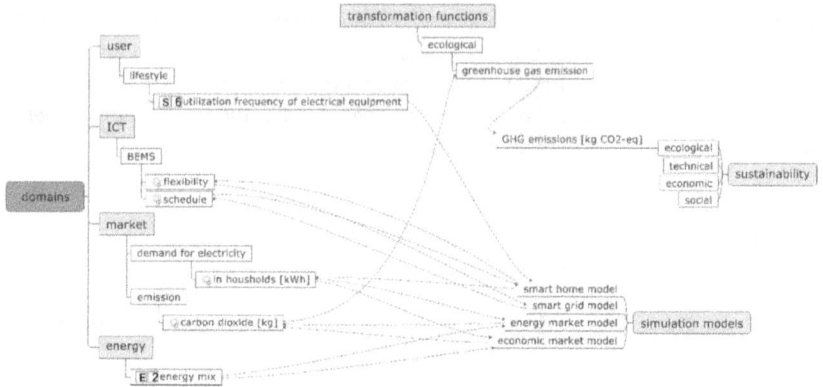

Figure 8: Example of the information model [17]

This example illustrates how the information model can support modeling scenarios. The modeling can take place in a mind map following the described structure and aims to support discussions in an interdisciplinary project team. The content is serializable in standard formats. However, modeling a scenario with many attributes, dependencies, and data flows can be highly complex and therefore would benefit from assistance and automation. Thus, the information model is integrated with Semantic Web technologies [23]. The Semantic Web was developed to add semantic descriptions to the World Wide Web but can also be used to make knowledge machine readable and processable with the following technologies. RDF (Resource Description Framework) is the common format to store data in triples of subject, object, and predicate. OWL (Web Ontology Language) is the common language to describe ontologies, which describe knowledge based on concepts (classes) and their relationships (properties). For the development and usage of ontologies, the ontology editor Protégé is used. The SPARQL Protocol And RDF Query Language (SPARQL) is used to query the content of RDF files or triple stores.

To make the content of the information model machine readable for further processing it is transformed to RDF and the structure of the information model is defined by an OWL ontology as described in [24]. Thus, the information model can be used to assist in the development of a so-called *high-level scenario*. Modeling this high-level scenario means describing the goals of a simulation, identifying needed simulation models, modeling data flows, and identifying potential dependencies between simulation models. Especially, for answering interdisciplinary questions, the modeling can get very complex. The information model in NEDS contains 29 domain objects, 231 attributes, and 18 sustainability criteria (The complete information model of NEDS can be found online[2]). The domain objects and attributes are listed in Table 36 and the sustainability criteria are presented in Section 5.2. As this number of objects can get hard to handle manually, queries can assist the user to find the needed information. For example, queries can show attributes or sustainability criteria, which are not connected to anything and recommend connections (more details and examples were published in [25]). Thus, it helps to use the full potential of the modeled scenario. Additionally, in several domains ontologies exist with domain knowledge, which can be integrated, e.g., to reference existing terminology.

The content of the information model can be used to assist the user not only in developing a high-level scenario but also in the steps toward an executable simulation. Because multiple simulation models may be needed for the interdisciplinary simulation in the PDES, co-simulation provides promising functionalities. Co-simulation allows to couple simulation models, which can represent different domains and can be implemented in different programming languages and paradigms [26]. Thus, co-simulation suits perfectly for building an interdisciplinary simulation of future scenarios. The information model can be used to assist the user in the planning of such a co-simulation. A scenario planning process integrated directly in the co-simulation framework "mosaik" is under development and described in [27] (see Figure 9). The left-hand side of Figure 9 represents the already explained integration of the information model. To develop a concrete co-simulation based on the "high-level" scenario planning in the information model, a catalog for co-simulation components has been developed. It uses the Functional Mockup Interface (FMI) standard [28] for the definition of variables and is implemented in a Semantic Media Wiki (SMW) [29], which allows the integration of its content in SPARQL queries. With these queries, the information model and the catalog can be used to find suitable components to build a simulation scenario, which consists of the simulation models, their parametrization, and the data flows between them. Additionally, the information model can be used to identify dependencies between different simulation models or the parametrization. For example, attributes, which are used for the parametrization of several different simulation models, should be modeled in the information model to ensure the consistency in the parametrization.

---

[2] *https://www.neds-niedersachsen.de/fileadmin/neds/pdf/NEDS_Informationmodel.zip*

The information model can also be used to integrate the knowledge from the high-level scenario description in the evaluation in two ways. On the one hand, the dependencies to the SEC in the PDES are modeled and as shortly described in [17], the information model was used to automatically generate a relational database schema to store the relevant data. For every attribute, the assumptions and results for the different alternatives, scenarios, and years can be stored in this database. Thus, it can make the parametrization and data exchange more transparent. On the other hand, the information model can also be used for simulation in other contexts to model dependencies and data flows of simulation models and make the evaluation more transparent.

With the described use cases, the information model supports the PDES described in Section 4.1. It was used in NEDS to structure discussions and make the goals and data flows transparent. Furthermore, it enables larger interdisciplinary projects to apply it with larger numbers of simulation models to handle complexity.

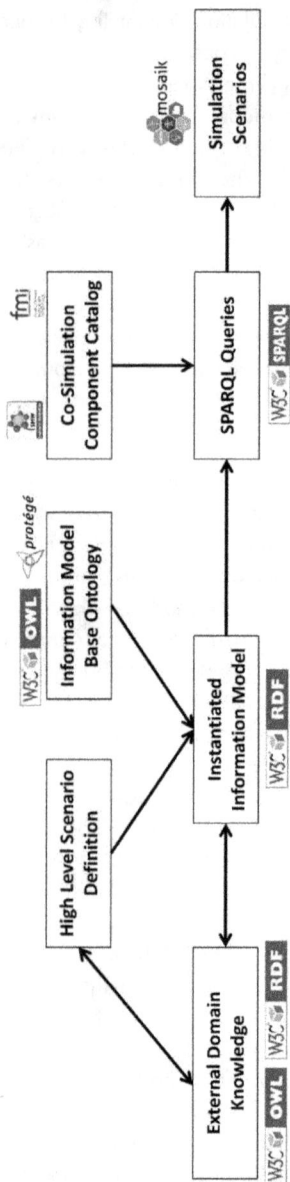

Figure 9: Overview of simulation planning process described in [27]

5. Sustainability as Evaluation Concept

F. Wille

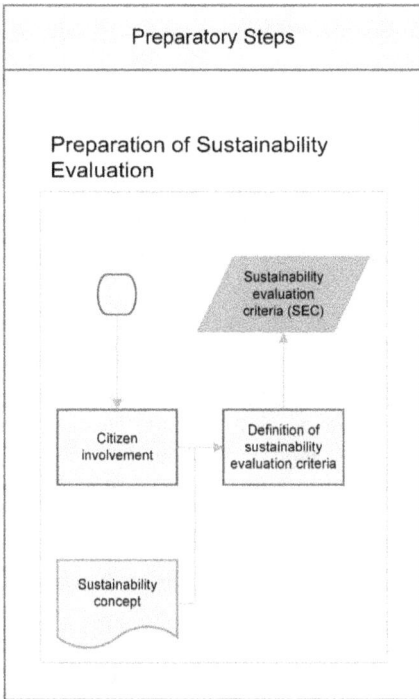

Figure 10: Excerpt from the PDES for defining relevant evaluation criteria

Defining and operationalizing the concept of sustainability is one of the first preparatory steps of the methodology for integrated development and evaluation of energy scenarios because this operationalization serves as a basis and provides input for many subsequent steps in the methodology. Figure 10 provides an overview of the process steps that lead to the operationalization of sustainability in terms of concrete, measurable criteria.

There is a myriad of possible criteria to evaluate an energy system. In our project, the focus for selecting criteria is their correspondence with the concept of the German term 'Nachhaltigkeit', which translates to the term 'sustainability', but is sometimes used with different connotations [30]. Originally, the term sustainability was employed in relation to forestry in 1713 by Hans Carl von Carlowitz [31] and developed further into a principle describing the usage of forest resources in a way, which would allow future generations equal or more usage of forest resources (Hartig, cited from [32]). At first, the sustainability principle was only applied to upholding limited forest resources, but it was expanded to incorporate other aspects which were to be upheld, e.g., biodiversity or the functional integrity of an ecosystem [32]. Its expansion to natural resources in a broad sense seems still consensus in concepts of sustainability [33]. This principle can be identified for example in the idea of dividing energy resources in renewable and non-renewable sources of energy. The current use of the term renewable energy as defined by the International Energy Agency as "energy that is derived from natural processes (e.g., sunlight and wind) that are replenished at a higher rate than they are consumed" [34] is relatively consistent within the literature [3] and often includes enumerations of wind, solar, geothermal, hydro, and tidal power and biomass as examples of renewable energy [3]. Coal, gas, oil, and uranium are considered non-renewable sources [35]. In our project, we emphasize this idea by evaluating the proportion of renewable energy resources in possible future energy systems. Even though the concepts of renewable energy and sustainability are strongly linked in public discourse, for achieving climate change goals, [3] argue that it could be detrimental to equate both concepts and that specifications of sustainability criteria are important for an evaluation of an energy system. Another notion is the long-term perspective of sustainability by considering effects for future generations, which according to Seghezzo [36] is neglected as part of the contemporary debate of sustainability but is nonetheless part of most ideas about sustainability in terms of intergenerational justice [33]. Long-term effects of the transition toward higher shares of energy from renewable sources can, for example, be analyzed with long-term energy scenarios.

According to Hellbrück and Kals [30], sustainability became a popular guiding principle for discussing environmental problems and necessary behavior changes with the Club of Rome Report "The limits to growth" in 1972. Having undergone changes in meaning since then, a common notion concerning the definition of sustainability appears to be its description by integrating goals within three different dimensions: environmental, economic and social [30]. This three-dimensional perspective was developed by the Enquête-commission as a guiding principle for a sustainable society regarding the three dimensions as equal to each other [37]. Due to the important influence of technical innovations on living standards and global developments, e.g., in information- and communication systems and especially in the field of renewable energies [37], it is appropriate to integrate as the fourth dimension of sustainability a technical perspective, which was also done by others before [38].

Another aspect of defining sustainability is its conception as strong or weak. The differentiation focuses on the degree of substitutability of natural capital [33]. From an economic perspective, different types of individual and collective legacies for future generations are associated with building, sustaining and reproducing capital stock, which can be categorized into physical capital, natural capital, cultivated natural capital (e.g., utilized agricultural area), social capital (e.g., institutions), human capital (e.g., education) and knowledge capital [33]. While in conceptions of strong sustainability, the idea is to keep the natural capital constant over time, in weak sustainability conceptions natural capital can be substituted by physical capital, as long as an aggregated utility value in terms of human welfare does not decline [33]. In our evaluation process, we do not define limits in criteria for substitutability, but compensation effects are taken into account when specifying the MCDA-method (see Section 11).

The sustainability evaluation criteria (SEC) are a significant output of the preparatory steps within the evaluation process as depicted in Figure 10. They are based on the analysis of empirical data from citizen involvement, which is described in Section 5.1, and the definition of SEC, which is described in Section 5.2. Both the empirical analysis as well as the definition of SEC build on the described concept of sustainability.

5.1 Empirical Analyses of Evaluation Criteria for a Sustainable Energy System and the Importance of Sustainability Dimensions

F. Wille

In order to realize a future energy system, societal acceptance and legitimization are of importance [39]. Even though there is enough literature from which to choose evaluation criteria for an energy system, we decided to specifically address the Lower Saxony context of this project and to analyze preference and evaluation patterns for a sustainable energy system in Lower Saxony. The aim is to develop a process, which allows integrating evaluation criteria and preferences as stated by groups of interest. Since there is no *a priori* reason to assume differences in the preference patterns of groups labeled "Lower Saxons" and "other states", our selection of groups of interest did not differentiate this aspect. Instead, we focused on acquiring two relatively homogenous groups to analyze their preferences toward the evaluation object of an energy system for Lower Saxony in 2050. Research questions are, first, which social, economic, environmental and technical evaluation criteria for a sustainable energy system in 2050 in Lower Saxony are verbally stated for those sustainability dimensions by professional experts and by thematically highly interested citizens and, second, how do these groups weigh the importance of the different sustainability dimensions for a sustainable energy system in 2050 in Lower Saxony?

Four *experts* were selected by the project consortium based on their professional expertise. The representative for the *economic dimension* works for the Institute for The World Economy, the representative for the *social dimension* for the German Trade Union Confederation (area of expertise energy politics), the representative for the *environmental dimension* for the Federal Agency for Nature Conservation (area of expertise nature conservation and renewable energies), and the representative for the *technical dimension* works for the Bundesnetzagentur (area of expertise energy regulation). Data material from this expert group was collected at a public symposium at the beginning of the project from (1) speeches delivered, (2) a moderated focus group discussion between the experts and (3) an open discussion with symposium participants. The produced speeches were based on an open question "According to your opinion, what features should a sustainable energy system have? What are your ideas for a future energy system in Lower Saxony?" (translated from German)[1].

The group of *thematically highly interested citizens* was recruited from a pool of 29 symposium participants who answered open questionnaire items ("What means sustainability for you?" and "What are the important criteria of a sustainable energy system for you?") and indicated willingness to participate in semi-structured interviews of approximately one hour by separately leaving contact information. Ten interviews were conducted with participants between 27 and 75 years old.

A qualitative content analysis according to Mayring [40] was conducted. The analysis was performed with *MAXQDA18* software. Sustainability dimensions were coded deductively and sustainability evaluation criteria were coded inductively based on data material from experts as well as highly interested citizens. Interrater reliabilities lie between *kappa* = .76 and *kappa* = .97 [41] for the different types of data material.

To answer the first research question, the inductively coded evaluation criteria were summarized for the group of experts and highly interested citizens, broken down into the different dimensions of sustainability, which were coded deductively in the first step of the analysis. The developed evaluation criteria are the result of inductive category building according to Mayring's technique of a summarizing content analysis [40], which builds its' categories via paraphrasing text segments, generalizing paraphrases and abstraction. In this way, a category system with different levels of abstraction (for example, categories at the top level and below that, sub-categories) can be built. Depending on the sustainability dimension, our category system encompasses in descending order of abstraction *main categories*, *upper categories* and *sub-categories*. The sub-categories designate the evaluation criteria. In cases, where a sub-category was not coded to belong to an upper category, it stands alone in the following visualizations. To this logic, we made one exception for the case of the upper category *acceptance*, because even though it only contains the one sub-category consider acceptance, it was stated so often, that we wanted to relate this by an upper category, even if the paraphrased statements were not diverse enough in content to build different "acceptance sub-categories".

For an aggregated overview of the category system with its evaluation criteria, see Figures 10, 11, 12 and 13.

---

[1] All data material, instructions, coding schemes and results are originally in German. For reporting, important parts are translated.

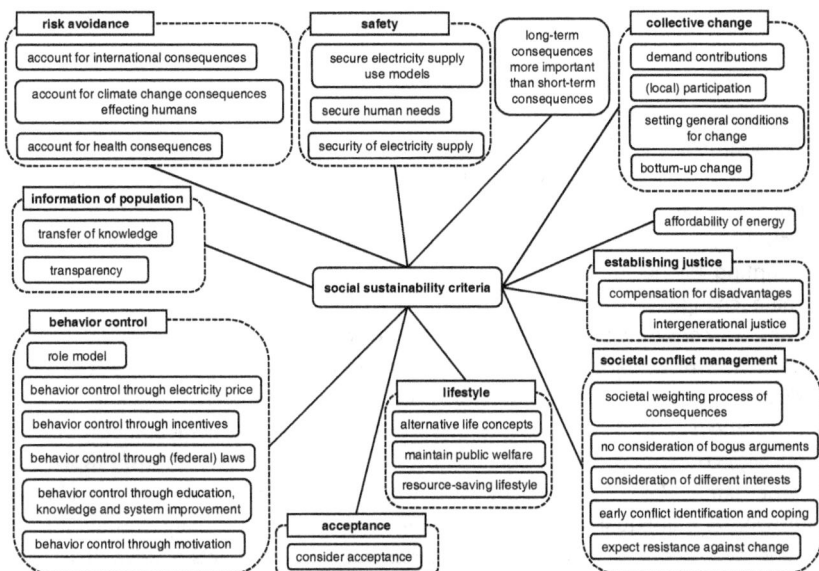

Figure 11: Social sustainability criteria from qualitative context analysis

As can be seen, most criteria were coded for the social dimension of sustainability (31 criteria). Eleven upper categories were built:

The criteria (local) participation, demand contributions, setting general conditions for change and bottom-up change were grouped as part of an upper category *collective change*. They aim at integrating humans with different roles in regard to the energy system (e.g., citizen, entrepreneur, politician) in the transformation process. *Acceptance* as upper category describes the demand to consider acceptance in society concerning planning and implementation of energy systems. The upper category *information of population* includes aspects of transference of knowledge and education as well as transparency of the process. Transparency in this context was sometimes described as a factor influencing acceptance directly or indirectly via (local) participation. The upper category *behavior control* summarizes discussed possibilities and ideas on how to influence behavior to achieve a sustainable energy system. This includes the use of and being a role model, regulation of electricity pricing, incentivizing sustainable choices, regulation through (federal) laws and motivational interventions as well as education, knowledge and system improvement. The upper category *safety* as part of the social sustainability dimension encompasses demanding of a technical energy system to secure and focus on certain human needs among which supply with electricity and a secure use model or use pattern of electricity were further distinguishable. Evaluations of whether the use patterns should change to be compatible with a transformed energy system or if the new system should be built to minimize changes in use patterns and secure maximum freedom in electricity use, varied. The criterion resource-saving lifestyle in the upper category *lifestyle* is comparable to secure electricity use models, but does not address electricity use and states the need to change current lifestyles. *Lifestyle* also includes maintaining public welfare and living standards, as well as discussions of alternative living concepts with for example higher flexibility in working hours. The upper category *risk avoidance* comprises accounting for international consequences of for example resource conflicts, accounting for health consequences (e.g., respiration, smog) and accounting for climate change consequences affecting humans. The criteria societal weighting process of consequences, no consideration of bogus arguments, consideration of different interests, early conflict identification and coping as well as expectancy of resistance against change describe an approach toward *societal conflict management* in which perspectives, arguments, and interests of different groups are (critically) considered. The upper category *establishing justice* comprises aspects of compensation for disadvantages and intergenerational justice, which aim at compensating unequal distributions of advantages and disadvantages resulting from an energy system within and between generations. Such aspects of advantages and disadvantages are also discussed in terms of the criterion affordability of energy. The criterion long-term consequences are more important than short-term consequences was stated only once and was sorted to the social dimension because it can be viewed as close to a societal weighting process of consequences, but an interpretation as a statement for a general weighting principle of consequences of an energy system is also possible.

Statements coded as belonging to the environmental sustainability dimension (18 criteria) are summarized in following upper categories: *environmental compatibility, minimize emissions* and *minimize impacts,* which are grouped as part of a main category *environmental protection,* to which the criterion climate protection can be sorted as well, while *maintenance and restoration of nature* and *implementation of nature protection goals* were grouped within a main category *nature conservancy,* as was the criterion species protection. The upper category *follow ecological boundary conditions* describes the demand of maintaining an ecological balance and observing ecological limits, which also can be interpreted as a statement toward weighting of criteria. In demarcation to the social criterion account for climate change consequences affecting humans, where the focus lies in explicitly stating the relevance for humans, the environmental criteria consider catastrophic climate impacts describes climate impacts without relation to effects on humans and was thus coded as part of environmental sustainability criteria.

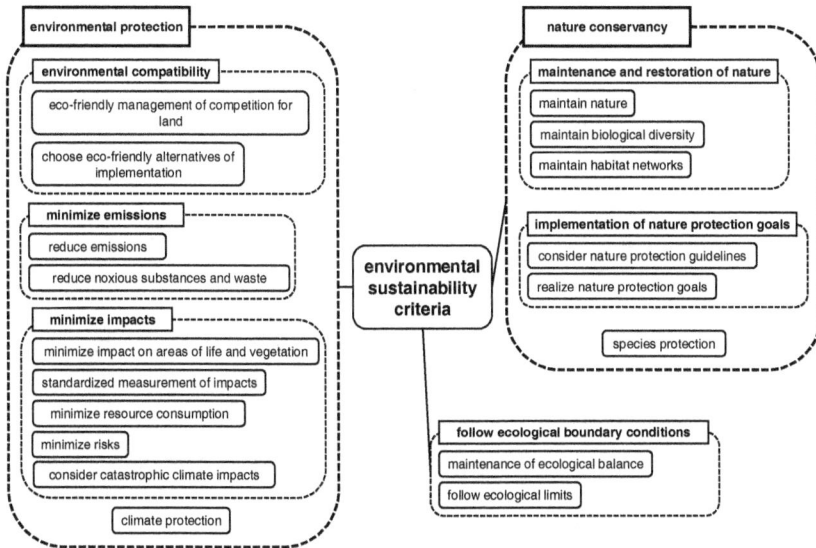

Figure 12: Environmental sustainability criteria from qualitative context analysis

Most economic sustainability criteria (13) are grouped in the upper category *weighing up of costs*, which encompasses the idea of using cost analyses to evaluate change options, to choose lower-cost options and to integrate aspects like consequences of interventions or environmental changes in a cost calculation. The differentiation between short-term and long-term consequences arises also in this context, but without a preference statement toward weighting. *Competition* as upper category comprises the existence of a competitive market economy as a regulating instrument, the maintenance of (international) competitiveness of Lower Saxony's industry and the goal to attain competitiveness of renewables. The upper category *preservation of work capacity* includes the demand to consider loss and gain of workplaces in economic decisions, to support the adaption of workers to new workplaces and to plan and design adequate workplaces in the energy sector. Since the focus lies on maintaining a work-force, it was categorized as belonging to economic sustainability criteria. The criterion preservation of structure and structural change encompasses the idea to maintain closed industrial value chains and the use of regional resources in spite of structural changes in the energy system. The criterion alternative economic system was stated by the group of interested citizens, who doubted the sustainability of growth-oriented economic systems.

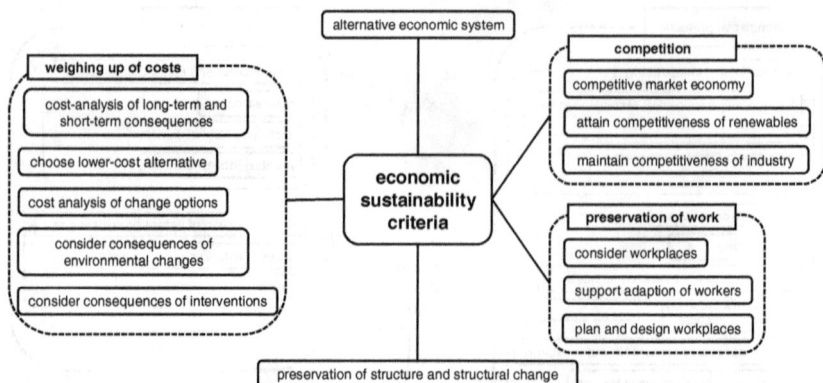

Figure 13: Economic sustainability criteria from qualitative context analysis

Since technical sustainability criteria are less known, we describe a few examples from paraphrased text segments for illustrative purposes. Criteria for technical sustainability (twelve) are reliable systems (e.g., low susceptibility of systems, need of storage system for reliable supply, grid stability), intelligent systems (e.g., information and communication technology to better control energy distribution), flexible systems (e.g., adaptiveness, sector coupling, no preference for certain types of power plants), safe systems (e.g., operational safety, too high safety risks concerning atomic disposal and accidents), efficient systems, and frugal systems (e.g., improve energy savings, consider whole life-cycle of technologies, use of existing infrastructure). The upper category *long-lasting systems* encompass aspects such as reusability of materials in the manufacturing process, long operability of energy technologies due to good mending and maintenance possibilities and use of regenerating energy resources in contrast to those which do not regenerate. The criterion advancement of technologies comprises statements toward promoting the development of storage systems and promoting research without constraints on the type of technology. A *systems' closeness to people* is characterized by its decentralization, adaptiveness to human needs and the use of local potentials for energy conversion.

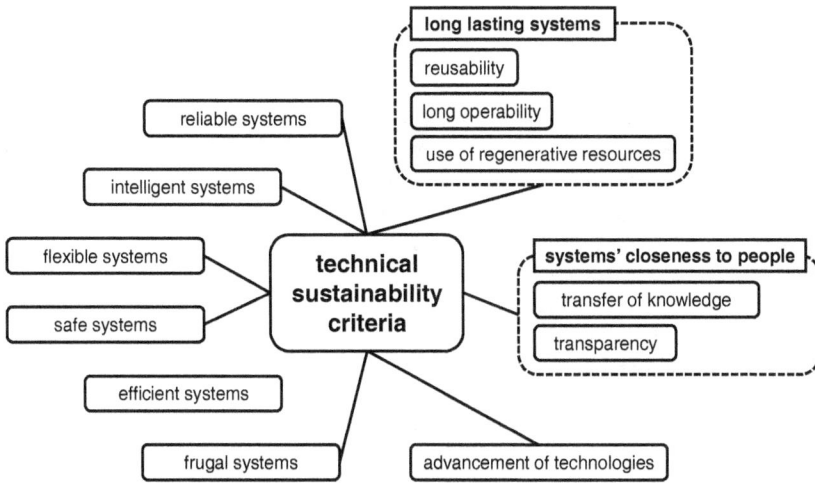

Figure 14: Technical sustainability criteria from qualitative context analysis

The derived criteria for our four sustainability dimensions can be related to existing criteria. The extracted social sustainability criteria correspond with topics of social indicator sets developed in a similar European project (NEDS), which aimed at devising an indicator set of social sustainability criteria for energy systems [42]. In that project, four main criteria are derived: "political stability and legitimacy", "security and reliability of energy provision", "social and individual risks" and "quality of life". Values of a social sustainability concept described by Hull [43] such as equality, work, liberty, social solidarity, universal access to basic goods, intergenerational justice, and social justice are also discussed by our group of experts and interested citizens. Furthermore, principles such as political participation, which already appear in the 'Brundtland Report' (WCED; [44]), are stated and were grouped in the upper category of *collective change*. A relatively large amount of codes is grouped in the upper category *behavior control,* which puts an emphasis on possibilities of behavior change methods to achieve a sustainable energy system. This category is not suited for operationalizing social sustainability in this current project, because the evaluation object consists of different states of systems and the category focusses on the process of how to achieve a certain system state. However, since questions of behavior change are viewed as an integral part of a sustainability discourse [45], we included it in our social sustainability criteria. Environmental criteria, such as resource consumption, climate protection, biodiversity, ecotoxicity, air pollution, waste reduction and minimizing risks are similar to the criteria topics of the NEEDS project [46] and also with other literature there seems to be a high topical similarity (e.g., [47] [48]). The NEEDS project [46] specified thematically comparable economic sustainability criteria in terms of preservation of work capacity (direct jobs: "as the amount of employment directly related to building and operating the generating technology"), but the considered or weighed costs are less broad. They are restricted to operational costs and financial risks of the utility manager, while our content analysis identifies criteria like economic cost of consequences of environmental impacts and interventions. This might be reflective of what Hull [43, pp. 76-77] identifies as one of three basic approaches in the discourse on conceptualizing sustainable development, which he distinguishes according to their main ideas and values. What he terms "conservational neo-liberal economy" is described as "an idea of sustainable development that leads to economic growth taking account of natural and social limitations", which tries to value and factor in all environmental goods, as well as environmental effects of human activity into market mechanisms. Often, a technical dimension is not included in the conceptualizations of sustainability (e.g., [49, 50, 51]). Some of the stated technical criteria like reliability and safety are sometimes part of a social dimension as in safety and reliability of energy supply, whereas criteria such as frugality, flexibility and efficiency are often sorted with an economic dimension [42, 46]. Inconsistencies in the allocation of criteria to different dimensions are to be expected as a result of introducing this new sustainability dimension. An interesting consequence of this conceptualization is, however, that relatively new evaluation criteria such as human relatedness of a technical system are discussed. Comparing our empirical results on sustainability criteria with other criteria as outlined above, we find a large overlap in discussed topics. Thus, verbal statements from ten interviewed citizens, four public speaking experts from different areas of expertise and 29 questionnaire answers mainly reproduced topics already discussed as relevant aspects in the sustainability debate with the exception of the relatively newly introduced dimension of technical sustainability. This is a similar observation as a conclusion drawn by Hopwood et al. [50, p. 47]: "In most cases people bring to the debates on sustainable development already existing political and philosophical outlooks".

As part of multi-criteria decision analysis (MCDA), the importance of criteria for a certain group of decision-makers or stakeholders is often assessed. In our evaluation process, we do not have a specified group of decision makers. With our second research question, we thus aim at evaluating the preference ranking of the four sustainability dimensions as discussed by our group of experts and highly interested citizens.

One way to establish a preference ranking is to analyze verbal statements toward the weighting of different sustainability dimensions. This was done for a group of experts. Since, in total, there was a low amount of "ranking statements" in the expert group (17 out of 482 codes), from which to deduce a preference order, we included a specific question in the interviews to rank the sustainability dimensions according to importance. Both procedures deduce a preference order based on explicit verbal statements. Taken together, the statements by the expert group in their talks, the focus group and open discussion point toward a preference of the environmental dimension or an equal weighting of all dimensions. Aspects of the environmental dimension are described as general boundary conditions, which need to be upheld, and the other dimensions are to be weighed more or less equally depending on the speaker after that. To exemplify this summary, view the translated statements from experts one to four (for expert two, no weighting statement was coded) and an example from the focus group as well as an open discussion:

"From a nature conservation perspective not only the costs are important, but even more so a good cost-benefit ratio. And we have to take conflicts with other societal goals seriously and to this belongs also the nature conservation goals" (expert_1, 8)

"this has to happen of course within the ecological or planetary load limits" (expert_3, 7)

"One needs basically a strict upper limit of emissions, which one defines and if we take the politically set [...] two degrees goal, then we have a fixed amount of emissions we have to emit and there I believe, that we have to achieve that, to uphold this goal." (expert_4, 45)

"the equality of different aspects arranges of course beneath the upholding of planetary load limits somehow" (focus group, 36)

"insofar I find this 'it has to be economically driven' in fact a little [not understandable] too short-sighted because for me it is always the questions, what must I actually include in the calculation" (open discussion, 50)

The preference order, based on explicit statements for the group of interested citizens, is summarized in a dominance matrix in Table 1. In the group of interested citizens, the environmental dimension also ranks first, followed by the social, economic and technical dimensions.

Table 1: Dominance matrix of stated ranks in a group of citizens (row beats column)

	environ-mental	social	economic	technical	dominance (sum)
environmental	-	5	7	8	20
social	3	-	6	8	17
economic	3	4	-	5	12
technical	2	2	5	-	9

Interpreting this type of preference order indicator, which is based on explicit verbal statements, as the only possible indicator, would be too reduced because it neglects the situation, in which the statements were made, the function of verbal behavior and the method used to deduce the indicator. The dominance of the environmental dimension in verbal statements could be for example largely attributable to the sustainability topic of the project and function as a means to further status quo interests (e.g., [52]; [53]) or even other interests. To consider this possibility, a preference indicator, which is closer to observable behavior, such as the amount of produced verbal behavior on a certain sustainability dimension, can be evaluated. Since producing verbal behavior means energy expenditure for an organism, we assume the more energy an organism spends on a certain topic, the more importance that topic has, i.e., the higher the preference for that dimension. As such an implicit indicator of a preference order, we choose the number of codes per sustainability dimension. The results for the group of interested citizens and experts are displayed in Figure 15 and Figure 16.

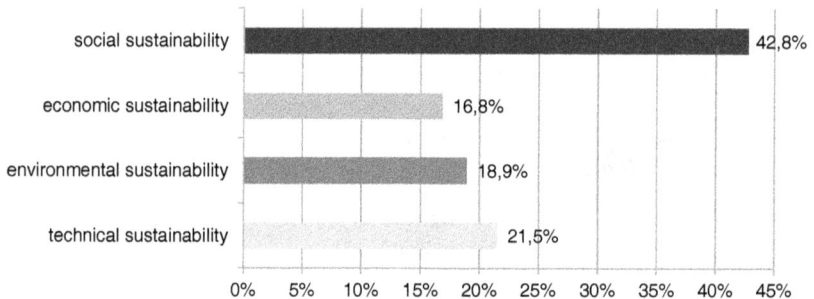

Figure 15: Code counts in percent for a group of highly interested citizens (interviews and questionnaires)

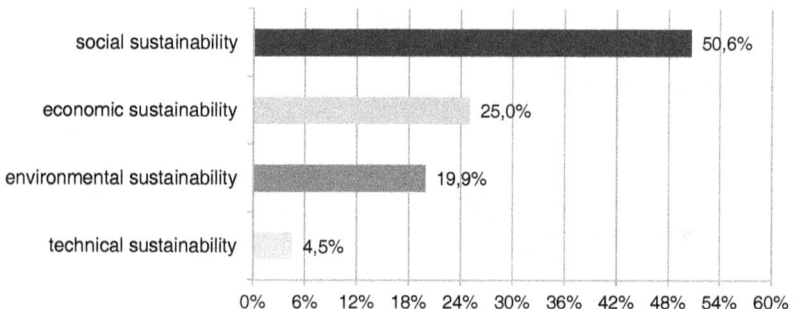

Figure 16: Code counts in percent for a group of experts(talks and focus group discussion)

For both groups, the social sustainability dimension is the one with the most codes, followed for the group of interested citizens by the technical, environmental and economic sustainability dimensions with percentages ranging slightly between 16.8% and 21.5%. For the groups of experts, the social sustainability dimension is followed in code counts by economic and environmental sustainability (25% and 19.9% respectively) and with a larger difference by the technical dimension (4.5%). This might be due to the relative newness of incorporating such a dimension in a conceptualization of sustainability and the fact, that in contrast to the interviewees, experts' statements were open in comparison to semi-structured. As can be seen in Figure 17 even in relation to their dimension of expertise, for each speaker, the talks consisted of mainly or to an equal amount of social sustainability codes.

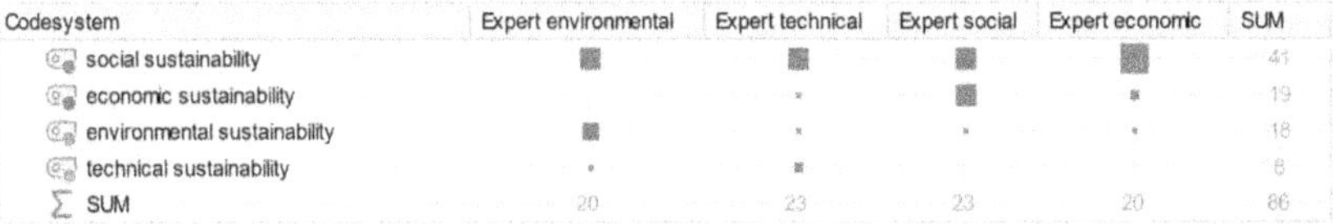

Codesystem	Expert environmental	Expert technical	Expert social	Expert economic	SUM
social sustainability	▓	▓	▓	▓	41
economic sustainability		·	▓	▪	19
environmental sustainability	▓	·	·	·	18
technical sustainability	·	▪			8
Σ SUM	20	23	23	20	86

Figure 17: Code counts for a group of experts as representatives of their sustainability dimension (only talks, the symbol size is relative to total count)

Depending on the indicator chosen, different weighting options for the sustainability dimensions in the MCDA analysis are possible. Based on explicit verbal statements, one can rank the environmental dimension first and the remaining three dimensions equally for the two groups (assigning percentage values as weights for the dimensions based on the dominance matrix or code counts would suggest a precision of weighing decisions not justified by the method of analysis). Grounding preference analysis on explicit verbal behavior, be it in forms of qualitative analysis, questionnaires or through moderated workshop procedures (as often done in an MCDA process) is generally viewed as legitimate [54]. Considering the very different weighting outcome, when basing the indicator on implicit verbal behavior (rank the social sustainability dimension first and then the other dimensions equally), one should question this approach as means to deduce preference information. For a preference analysis, it is important to reflect on the group of interest as well as on chosen indicators. Since the explicit weighting results are probably explicable in terms of situational factors, choosing an implicit behavioral indicator might be a better indicator of preference.

Relating the results of deduced evaluation criteria and weighting of sustainability dimensions of a group of experts and a group of highly interested citizens to the concept of sustainability, two conclusions for the overall method development (see Section 4) are important: First, we analyzed two groups, which reproduced a large number of criteria discussed within the sustainability debate. New criteria and ideas were mainly identified for the dimension of technical sustainability. Trying to include a wide range of criteria from this analysis as well as from existing literature is thus a good procedure, especially if trying to integrate a new concept such as technical sustainability. Second, even though it is possible to deduce a meaningful preference ranking for the sustainability dimensions, in light of the concept of sustainability itself, an equal weighting of all dimensions instead of representing a selection of different group preferences is preferable for this level of decision making. As Hull [43] points out, there is no consensus on how the idea of sustainability should be construed. Even though a sustainability principle – formulated in a broad way as bringing three fundamental spheres (economic, social, natural) of human existence and activity into an equilibrium – gained wide acceptance, there are still in everyday life a multitude of perspectives, which are all justified. Before advancing to a concrete implementation of energy system changes, which needs conflict and interest management, which was an important part of the social sustainability dimension, an equal weighting of interests and preferences is in best accordance with our concept of sustainability. When all perspectives are to be respected, a procedure, which allows evaluating different weightings and the consequences associated with them, seems useful for decision problems, in which a large number of different groups and perspectives is included. In this way, the results and consequences of different weighting decisions can be made transparent.

5.2 Synthesis of Evaluation Criteria

M. Dumeier, F. Wille, T. Witt

The abstractness of the identified criteria necessitates a further operationalization applicable to our specific evaluation object. Our selection of evaluation criteria is guided by the principle of incorporating a broad spectrum of criteria from every sustainability dimension identified in the empirical analysis, although it is not possible to integrate all deduced sustainability criteria into the MCDA analysis. We also collected criteria from the literature [55, 38] and then condensed, and arranged in a hierarchy a set of potentially relevant evaluation criteria. The selection is restricted by the availability of indicators from our simulation models. Thus, the final criteria catalog is a synthesis from the literature, the results of the qualitative content analysis, and the available attributes in the optimization and simulation models (see Section 9). Table 2 depicts the criteria hierarchy derived for the evaluation of Lower Saxony's power system until 2050. The major objective is split into four sub-objectives according to the four sustainability dimensions, which are further broken down into several measurable criteria. The number of criteria per sub-objective varies from two to seven, which needs to be considered when the criteria are weighed for the MCDA. Further information on how the performance scores for each alternative and criterion are calculated is provided in Section 9. Further information on how the performance scores are aggregated with the MCDA method PROMETHEE is described in Section 11.

Table 2: Criteria hierarchy for evaluating the sustainability of Lower Saxony's power system

Sub-Objective	Criteria	Unit
Technical	Percentage of plants utilizing renewable energies	%
	Grid efficiency	share of output %
Social	Import quota for energy sources used	%
	The ratio of wage to capital income	%
	Share of expenditure on electricity of total consumption expenditure	%
	Behavioral adaptation costs	€/capita
	Particulate matter formation	kg PM10-eq/MWh
	Photochemical oxidant formation	kg NMVOC/MWh
	Human toxicity	kg 1,4-DCB-eq/MWh
Environmental	Metal depletion	kg Fe-eq/MWh
	Fossil depletion	kg oil-eq/MWh
	Climate change	kg CO2-eq/MWh
	Terrestrial acidification	kg SO2-eq/MWh
	Freshwater eutrophication	kg P-eq/MWh
	Terrestrial ecotoxicity	kg 1,4-DCB-eq/MWh
	Agricultural land occupation	m²/ MWh
Economic	Real gross domestic product	1,000 €/capita
	Costs for electricity production and grid expansion	€/MWh

Note that the assignment of criteria to the different sub-objectives in Table 2 slightly differs from assignments in the literature (see, e.g., [38, 55]): Criteria, which are usually measured with a Life-Cycle Assessment (LCA), are usually assigned to the environmental sub-objective only. However, we split these criteria into two groups, to better take into account the results of the empirical analysis described in Section 5.1: If criteria have a *direct impact on human health*, they are assigned to the social sub-objective. For example, the exposure to the chemicals measured by the criteria is shown to cause a variety of negative health effects and thereby reduces life expectancy. If the criteria have a *direct impact on the environment*, e.g., through the contamination of freshwater or soil, which leads to a decreased biodiversity or to the consumption of finite resources, they are assigned to the environmental sub-objective.

6.   Development of Future Scenarios

J. S. Schwarz

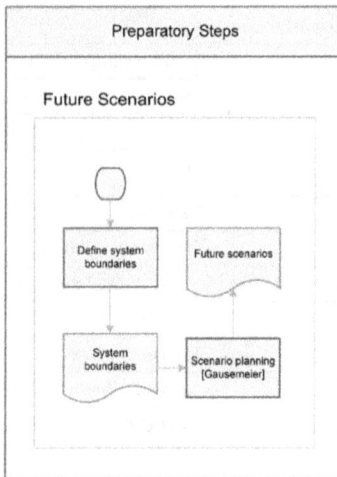

Figure 16: Development of future scenarios in the PDES

The development of future scenarios is located in the *Preparatory Steps* phase of the PDES (see Section 4.1) in parallel to the preparation of sustainability evaluation. In this context, a future scenario is a qualitative storyline of future developments, e.g., regarding the future energy system and its environment, which provides a common ground for the investigation of an interdisciplinary question. In a later stage of the PDES (see Section 8), the scenarios are used as the basis for quantifying those assumptions that are necessary for quantitative modeling (see Section 9).

The first task for the development of future scenarios is the definition of system boundaries for the modeled energy system, according to the general objectives. Typical system boundaries in energy scenarios are temporal, spatial, and energy sector-related [7]. In NEDS, the system boundary is defined as the power supply system (energy-sector related boundary) in Lower Saxony (spatial boundary), up to the year 2050 (temporal boundary).

In the next step, qualitative future scenarios are developed. For this development, the scenario planning process described by Gausemeier [19, 56] was used in NEDS and is described in Section 6.1. The resulting future scenarios are outlined in Section 6.2.

6.1   Scenario Planning

J. S. Schwarz

Scenario planning[1] is an expert-based management tool, which originates from strategic planning in companies in the 1970s [57, 58]. It is based on the two main principles systems thinking and multiple futures. System thinking addresses the

---

[1] Also called scenario management.

dependence between different influence factors in a complex field for enterprises. The term 'multiple futures' means thinking about multiple possible developments in the future, instead of only one prediction, to be prepared for changing circumstances. The process of scenario planning and its background is described in detail in [19] and [56]. It consists of the steps Scenario Preparation, Scenario-Field Analysis, Scenario Prognostic, Scenario Development, and Scenario Transfer. An overview of the scenario planning process is shown in Figure 19. Scenario Transfer is omitted because it was not done in NEDS as described by Gausemeier. This phase aims at developing strategies based on the scenarios and in [56] multiple qualitative processes are described to predict the probability of different scenarios and to deal with the opportunities and risks related to the different scenarios. In NEDS, potential strategies were developed based on quantitative simulation and evaluation.

Scenario planning provides a transparent process for collaboration in an interdisciplinary project team of domain experts. Within this process, the experts discuss the most important factors and possible developments to develop consistent future scenarios. The process provides also tools to support the experts in these tasks.

6.1.1 Scenario Preparation

The goal of scenario planning is to support the stakeholder(s) in making a decision. The starting point of the process is the definition of the *scenario field* and the *decision field*. The scenario field describes the scope of the scenarios, while the decision field describes the part of the

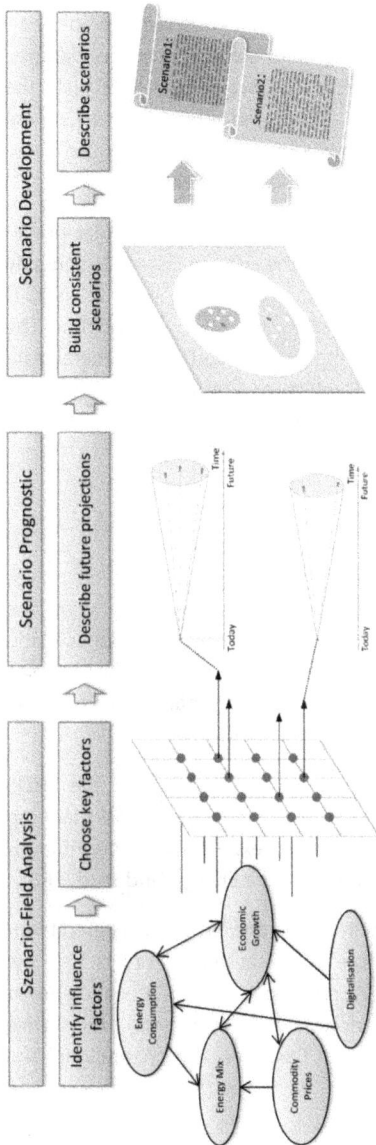

Figure 19: Scenario planning process (based on [56, p. 49])

scenarios, which can be influenced by the stakeholder(s). Based on this classification, three different types of future scenarios can be distinguished: External scenarios comprise only external factors and can be used to characterize the development of the environment of decision alternatives. In these scenarios, no decision field exists. Internal scenarios comprise only internal factors and can be used to describe possible decision alternatives. In these scenarios, the scenario field equals the decision field. Systems scenarios include both external and internal factors. These scenarios can be used to describe the development of complex systems. In these scenarios, the decision field is smaller compared to the scenario field.

The goal in NEDS is the development and evaluation of a sustainable future power supply with a focus on Lower Saxony. Because the power system is highly dependent on developments outside of Lower Saxony, the scenario field in NEDS is global with the considered domains energy, ICT, economy, user, and politics. The decision field is defined by the scope of action of the government of the federal state Lower Saxony. In NEDS, factors are considered, which can be influenced (internal factors), and factors, which cannot be influenced (external factors) by the stakeholders. Thus, the scenarios are system scenarios.

6.1.2 Scenario-Field Analysis

In this step, the relevant and characteristic influence factors for the scenario field are identified. Typically, this starts with brainstorming or brainwriting with the stakeholders to find influence factors. Having described the decision and scenario fields, the experts systematically identify the most relevant and characteristic influence factors – the so-called key factors (KF). For the identification, typically methods of influence analysis are used [59, 56].

For the analysis of the scenario field, a workshop with all partners of the NEDS project was organized. At this workshop, the first steps of the scenario planning process were executed. The process was supported by an expert, who brought in his experience from the scenario planning process in the Future Energy Grid study [60].

First, the participants collected influence factors in brainstorming and clustered them together. As results, 20 influence factors were found. For these influence factors, the participants filled out an influence matrix, which resulted in the system grid shown in Figure 20. It shows passive (being influenced by other factors) and active (influencing the other factors) influence of factors. Based on the system grid, the influence factors were discussed. Some of them were merged together in the discussion and finally, eleven KFs were identified.

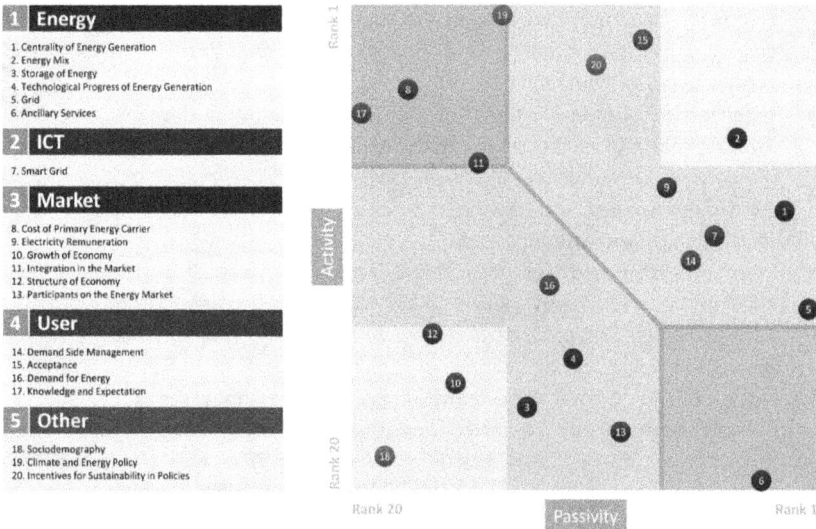

| **1** Energy |
| 1. Centrality of Energy Generation |
| 2. Energy Mix |
| 3. Storage of Energy |
| 4. Technological Progress of Energy Generation |
| 5. Grid |
| 6. Ancillary Services |

| **2** ICT |
| 7. Smart Grid |

| **3** Market |
| 8. Cost of Primary Energy Carrier |
| 9. Electricity Remuneration |
| 10. Growth of Economy |
| 11. Integration in the Market |
| 12. Structure of Economy |
| 13. Participants on the Energy Market |

| **4** User |
| 14. Demand Side Management |
| 15. Acceptance |
| 16. Demand for Energy |
| 17. Knowledge and Expectation |

| **5** Other |
| 18. Sociodemography |
| 19. Climate and Energy Policy |
| 20. Incentives for Sustainability in Policies |

Figure 20: System grid of influence factors

The following KFs were identified and described based on two dimensions of possible future development for the future scenarios in NEDS [61, 62]:

1. **Power grid:** This KF describes the nature of the power grid. The two dimensions are an *expansion of power lines (low/high)* and *use of controllable equipment (low/high)*.

2. **Topology of distributed energy resources (DER):** This KF represents the spatial distribution and size of DER. The two dimensions are the *unit size (small/large)* and *distance to consumers (near/far)*.

3. **Digitalization in the distribution grid:** This KF describes the level of digitalization in the distribution grid. The two dimensions are the *dissemination of smart meter gateways (low/high)* and *diffusion of ICT-infrastructure in the distribution grid (low/high)*.

4. **Energy management:** This KF describes the diffusion of energy management systems in private households and industry. The two dimensions are *application in private households (low/high)* and *application in the industry (low/high)*.

5. **Energy mix:** This KF describes the shares of renewable and fossil energies in the electricity generation. The two dimensions are the *share of renewable energies (low/high)* and the *share of fossil energies (low/high)*.

6. **Demand for energy (in households):** The energy demand of private households is dependent on the consumer behavior and number of electric devices per capita. The two dimensions are *diffusion of resource-saving behavior (low/high)* and *number of electric devices per capita (low/high)*.

7. **Economic structure:** This KF describes the energy intensity and growth of the economy. The two dimensions are the *economic growth rate (low/high)* and *energy intensity (low/high)*.

8. **Levelized costs of electricity:** This KF describes the future development of the levelized costs of electricity for both renewable and fossil energies. The two dimensions are *levelized costs of electricity of fossil-fueled power plants (decrease/increase)* and *levelized costs of electricity of renewable energy technologies (decrease/increase)*.

9. **Intentions of energy politics:** This KF describes national and international developments in energy policy. The two dimensions are *market regulation (create markets/strong regulation)* and *international coordination of the energy transition (low/high)*.

10. **Knowledge and perceived control:** This KF describes the knowledge of individuals about and perceived opportunities to control renewable energy technologies in the smart home or smart grid. The two dimensions are *knowledge about the environment (low/high)* and *perceived control (low/high)*.

11. **Acceptance:** This KF describes the acceptance of renewable energy technologies on both individual and societal levels as a function of cost-benefit ratios. The two dimensions are *individual cost-benefit ratio (low/high)* and *societal cost-benefit ratio (low/high)*.

6.1.3 Scenario Prognostic

For the identified KFs, projections are defined. The projections are possible developments up to a certain point in time to span a broad range of possible future developments. For every dimension, two different future developments were defined, so that every KF has altogether four projections as a combination of its two dimensions. These projections are described with a short text.

All the descriptions of the project NEDS can be found in [62] (in German). A detailed example of one KF's projections is depicted in Figure 21. It shows the KF *topology of DER* with the dimensions' *unit size* and *distance to the consumer*. The two dimensions are combined and build the following four projections:

Distributed and small power generating units (2A): This projection describes a future power system with a focus on small power generating units, e.g., single wind turbines, small biogas plants, and small ground-mounted and rooftop photovoltaic plants. The units are distributed and need power lines to transport their energy.

Distributed renewable power plants (2B): This projection describes a future power system with a focus on large power plants, e.g., ground-mounted photovoltaic parks, off- and onshore wind parks, large biogas plants, or fossil power plants. The plants are in distance to high demand areas and need power lines to transport their energy.

Local and small power generating units (2C): This projection describes a future power system with a focus on small power generating units, e.g., single wind turbines, small biogas plants, and small ground-mounted and rooftop photovoltaic plants. The generating units are near to consumers and the energy has not necessarily to be transported over long distances.

Conventional power plants (2D): This projection describes a future power system with a focus on large power plants, e.g., ground-mounted photovoltaic parks, off- and onshore wind parks, large biogas plants, or fossil power plants. The generating units are near to high demand areas and the energy has not necessarily to be transported over long distances.

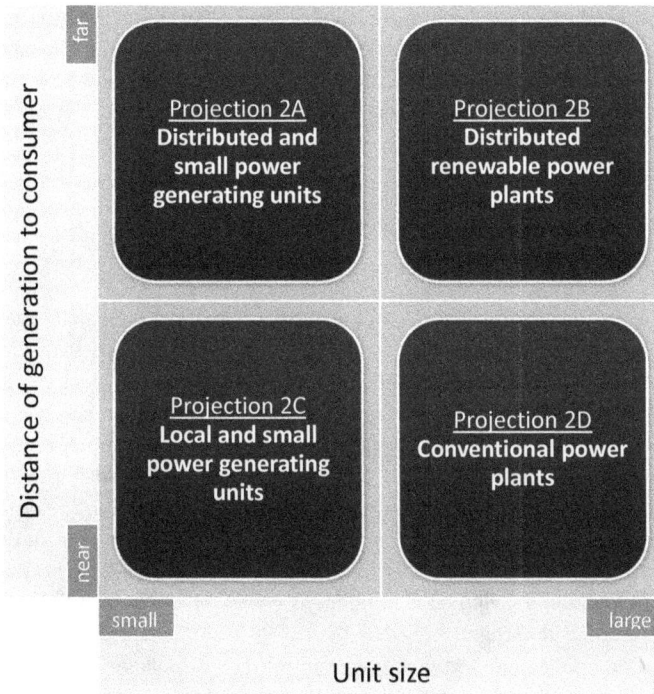

Figure 21: Projections of key factor "topology of distributed energy resources"

6.1.4 Scenario Development

The experts check the different projections for pair-wise consistency and the results are recorded in a consistency matrix. Based on this matrix, scenario software uses cluster analysis to build projection bundles, which represent consistent combinations of projections and thus possible future scenarios. In the last step, the domain experts write storylines, i.e., textual descriptions, for all future scenarios.

In NEDS, the project team filled out the consistency matrix based on the defined projections of the KFs. In contrast to the influence matrix, different ratings from different participants cannot automatically be merged. Thus, the matrix was subdivided and discussed in small groups.

Cluster analysis with the software Scenario Manager[2] was executed on the results and used to build five raw scenarios. The raw scenarios still contain uncertainty about some key factors' projections if the consistency matrix allowed no clear choice of a single projection but multiple were possible. Thus, the most suitable projections were chosen based on the discussion. Figure 22 shows these raw scenarios consisting of one projection for each KF. For each scenario, a storyline was developed, and meaningful titles were determined.

6.2   Description of Future Scenarios

J. Minnemann

Scenario 1 "smart consumption with economic growth" is characterized by a combination of flexible electricity demand and economic growth with increasing energy intensity. Fossil energy sources remain competitive in the long term, as ineffective promotion and lack of technical progress in renewable energies, among other things, hinder the transformation of the energy mix. The energy mix, therefore, remains largely unchanged, so that the topology of the distributed energy resources stays the same. Large grid extensions are not necessary. In the distribution grids, digitalization is progressing and is enabling a high degree of flexibility in generation and consumption in households and industry. Flexibility in households is supported by resource-saving consumption, as energy efficiency plays an important role. The resource-conserving behavior of society is supported by a high level of environmental knowledge and perceived control. The population is willing to take social risks, but the individual cost-benefit ratio is in the foreground.

---

[2] http://www.scmi.de/de/software/scenario-manager

#	Key factor	Scenario 1	Scenario 2	Scenario 3	Scenario 4	Scenario 5
1	Power Grid	Status quo	High expansion of power lines		Replacement of equipment and expansion of power lines	Intelligent Replacement of equipment
2	Topology of distributed energy resources	Conventional power generating units	Local and small power generating units	Distributed renewable power plants	Distributed and small power generating units	Local and small power generating units
3	Digitalization in the distribution grid	Widespread digitalization	Lost opportunity (missing ICT infrastructure)	No digitalization	Widespread digitalization	Widespread digitalization
4	Energy management	Widespread energy management	Widespread energy management	Small diffusion of energy management	No energy management in households	Widespread energy management
5	Energy mix	Status quo	Renewable energies	Fossil backup power plants	Renewable energies	Renewable energies
6	Demand for energy (in households)	Modern green	Sufficiency	Users focus on consumption		Modern green
7	Economic structure	Grey growth	Grey stagnation		Green growth	Green growth
8	Levelized costs of electricity	Fossil power plants are still competitive	Increasing costs for fossil power plants	Fossil power plants are still competitive	Cost advantage for renewable power plants	Cost advantage for renewable power plants
9	Intentions of energy politics	Little regulation of energy resources and no coordination in Europe	Regulation of energy resources and coordination in Europe	Regulation of energy resources and no coordination in Europe	Little regulation of energy resources and high coordination in Europe	Little regulation of energy resources and high coordination in Europe
10	Knowledge and perceived control	New technologies are used efficiently	New technologies are used efficiently	Stagnating usage of new technologies	Potential of new technologies is not used	New technologies are used efficiently
11	Acceptance	Opportunistic behavior	People support energy transition	Resistance		People support energy transition

Figure 22: Raw Scenarios with projections for each key factor

43

One reason for the stagnating economy is a sharp rise in fossil electricity production costs and stagnation in renewable electricity production costs. The population has a high level of environmental knowledge and is actively involved in the energy transition, for example through an efficient lifestyle and declining electricity consumption in households. A high diffusion of local small-scale distributed energy resources such as PV installations was achieved, which favors the spread of renewable energy production. A Europe-wide climate policy supports these developments. Both industrial and household customers have integrated active energy management on a large scale, but its potential cannot be fully exploited due to insufficient information and communication technology infrastructure. The expansion of power lines is therefore necessary.

"Competitive conventional power plants and untapped renewable potential" is the title of scenario 3. Although a high proportion of renewable power plants characterizes the scenario, the potential cannot be fully exploited due to a lack of infrastructure. Although distributed large-scale renewable power plants are being built, conventional backup power plants are necessary to maintain the stable operation of the grid. This result in high demand for grid expansion, but the distribution grids are little digitalized and demand flexibility has not been achieved. In the economy, energy intensity increases with the stagnation of growth. At the national level, there is a strong intervention in the energy markets and fossil fuels remain competitive. Consumers are more consumption-oriented. There is a stagnating implementation of savings potentials at the household level and in general, the population tends to resist more against renewable energies.

Scenario 4 is called "energy transition without the support of the population". In this scenario, the transition of the energy system is largely successful, although the population resists. Power plants are installed away from consumers and a high proportion of renewable energy sources can be achieved. The information and communication technology infrastructure in the distribution grids has been greatly expanded and investments are being made in intelligent resources in the grid. The industry can largely integrate energy management into its processes. Although the environmental knowledge of the population is high, the possibilities for control are perceived as limited. Despite better knowledge, the population is still consumer-oriented, and measures to make consumption more flexible in households are scarcely applied. Changes that are necessary for the course of the energy transition lead to resistance among the population. The electricity generation costs of fossil sources of energy are rising, while those of renewable sources of energy are falling. With simultaneous economic growth, the energy intensity in the production of goods and services decreases. Coordinated climate policy is successful at the European level.

The "cross-sectoral energy transition" succeeds in scenario 5, since in this scenario, conditions could be created for full support of the energy transition. An important part is the realization of local small-scale distributed energy resources. The energy mix is characterized by renewable energies, as these achieve a clear cost advantage in electricity generation. Digitalization of the distribution grid and maximum demand flexibility at both household and industrial levels have been achieved, but this has required a high exchange of operating resources in the grids. The economy is experiencing growth characterized by the production of less energy-intensive goods and services. A Europe-wide coordinated energy policy with little market regulation has been achieved. User behavior at the household level is ecologically motivated. There is an efficient application of renewable energy technologies, and the acceptance of the energy transition is high. The population is involved in the energy transition.

7. Diffusion and Adoption of Innovations for the Energy Transition

M. Kleinau, J. Minnemann, C. Busse

Amongst scientists, broad agreement exists that a fast energy transition is necessary for limiting the ongoing climate change. Given that numerous technological innovations are already available for facilitating the energy transition, more scholarly attention needs to turn to the processes of diffusion of these innovations [63] [64]. Diffusion as the macro-level process through which an innovation spreads through the population of available adopters [65] is apparently not taking place at the speed that would be required for slowing global warming. Macro-level innovation diffusion is the aggregate outcome of numerous micro-level innovation adoption processes of households, firms and governmental institutions [65]. Accordingly, this section seeks to answer the research question what influences the adoption and diffusion of relevant innovations for Lower Saxony's energy supply.

Our analysis of the created scenarios constitutes the starting point for the identification of crucial innovations. Our goal is to describe and analyze the diffusion of these crucial innovations, including the social systems consisting of parties such as (potential) adopters of energy-related innovations and change agents. With a thorough understanding of drivers and obstacles, sophisticated transformation strategies may be deduced and resources needed for the promotion of certain innovations can be allocated more efficiently. Furthermore, measures could be taken that aim at altering communication structures in said social systems, for example by increasing the visibility of successful adoptions or by adapting the communication between change agents and adopters. In addition to these more practical implications, another goal is to support the empirical foundation of the created scenarios, by providing forecasts for the diffusion of the selected innovations as well as adding arguments and explanations for possible diffusion paths.

The next Section 7.1 illuminates diffusion of innovation theory as the theoretical foundation for this study. The following Section 7.2 describes the process for choosing the most relevant innovations. Afterward, we sketch our research design for analyzing the potential for widespread adoption of crucial innovations in Section 7.3. A deeper understanding of influence factors and perspectives will be gained through qualitative interviews with both, adopters and change agents. As our main results, we summarize diffusion studies for each of the selected innovations (Section 7.4), before we discuss practical implications as well as future research opportunities in the concluding Section 7.5.

7.1 Theoretical Grounding of the Diffusion of Innovations

The foundation for our research is Rogers' (2003) seminal diffusion of innovations theory, which provides an empirically supported, highly generalizable theoretical basis for studying the relationships between the main elements in the diffusion of innovations. In accordance with our research goal, diffusion of innovations theory is used not only for explaining diffusion paths but also for predicting and possibly influencing diffusion by means of third-party interventions (e.g., [66]).

Rogers defines innovation as "an idea, practice, or object that is perceived as new by an individual or other unit of adoption" [65, p. 12]. He emphasizes the perceived degree of novelty of the innovation as the decisive criterion: "If an idea seems new to the individual, it is an innovation" [65, p. 12]. Innovations comprise, most importantly, new products, services or procedures, which differ substantially from existing solutions [67].

At the level of analysis of the innovation itself, five features have been identified as adoption determinants, namely the relative advantage, compatibility, complexity, trialability, and observability of the innovation [65]. The relative advantage describes to what extent an innovation is perceived to be more advantageous than existing solutions. Compatibility means that an innovation is consistent with an individual's values, experiences and needs. The complexity of an innovation is the only attribute with a negative impact on the diffusion. A complex innovation is difficult to understand or use by the potential adopter. Trialability gives the potential adopter an opportunity to try out an innovation and therefore helps to reduce the perceived risk. Observability means that the result of an innovation's adoption is visible to others.

Another main element impacting diffusion are the communication channels through which potential adopters hear about the innovation and its features. Those channels comprise mass media, interpersonal channels or interactive communication over the internet. Since most individuals do not rely on scientific studies to evaluate innovation and much rather build their opinion on subjective evaluations of others, it is crucial for the diffusion of innovation, that the communicating parties differ in certain attributes to some degree [65].

The third main aspect is the time passing over the innovation-decision process. This process aims at decreasing uncertainty by information seeking and -processing and consists of five main steps: knowledge, persuasion, decision, implementation, and confirmation. This usual sequence can occur in different orders and take different periods of time. For our diffusion analysis, we are aiming to gather data from adopters finding themselves in different steps during this process in order to create a better understanding of the importance of these steps for the different innovations.

Still, the results focus on the adopter's way to his decision, rather than the implementation and confirmation. Moreover, each individual might have a different degree of innovativeness, which describes whether a person is relatively early in adopting innovations. The members of a social system differ in their degree of innovativeness and can be classified as innovators, early adopters, early majority, late majority, and laggards. Mapping the rate of adoptions for different points in time leads to an S-shaped curve [65].

The last main element is the social system in which an innovation diffuses. Its structure, norms and of course members affect the diffusion. In addition to the adopters, relevant actors within this system comprise opinion leaders, who provide information and advice about innovation to many other individuals in the system and change agents, who influence their client's innovation-decision process in a direction desired by his change agency [65].

It always needs to be considered that the adoption or rejection of innovation might lead to changes to the individual or the social system. These consequences can be classified in desirable or undesirable, direct or indirect and anticipated or unanticipated. Change agents may find themselves encountering undesirable consequences whenever they fail to anticipate important indirect effects associated with the adoption of an innovation [65]. Since such consequences appear on a higher level for the entire energy system, we will not portray them in this section, but later on in Section 11.

7.2 Selection of Innovations

Our diffusion analysis begins with the choice of relevant innovations. Therefore, it is necessary to identify those innovations that are relevant for the energy transition in Lower Saxony. We applied a *seven-step process* to identify these innovations, their adopters and change agents. The process, shown in Figure 23, consists of an evaluation of the detailed description of the future scenarios as well as scientific studies, identification of relevant actors within the diffusion process and a categorization of these actors in adopters and change agents. Subsequently, the process is described on a step-by-step basis.

Step 1
Identification of Innovations through Evaluation of Future Scenarios

Step 2
Identification of Innovations from Transformation Studies

Step 3
Selection of Innovations for Analysis

Step 4
Identification of Relevant Actors

Step 5
Determination of Adopters

Step 6
Determination of Change Agents

Step 7
Elaboration of a Synoptical Table

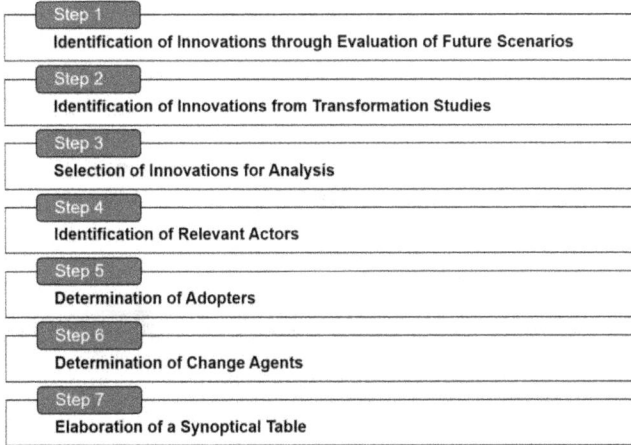

Figure 23: Process to identify innovations and relevant actors

The process of innovation identification as *step one* started with an evaluation of the detailed description of the future scenarios (see Section 6.2 for more information). Each scenario is based on eleven key factors with several different projections. The different projections of key factors were the starting point to identify innovations. Therefore, each scenario was analyzed in detail and possible innovations were selected. As a result, the first list of innovations was compiled. *Step two* extended this list with innovations found in five main transformation studies [68, 69, 70, 71, 72], resulting in 36 innovations being considered. Within *step three*, the list was adjusted by deleting those innovations for which an analysis of diffusion could not be done satisfactorily as the number of potential adopters is too low, the innovation itself is no product or the innovation only acts a tool. For example, all technological innovations regarding the transmission grid were deleted, as there are only four providers, with only one of them operating in Lower Saxony. An analysis of diffusion with such a small number of adopters does not make sense. Some others like load or energy management are not considered as they represent behavioral attitudes, which can be present to differing degrees, rather than products whose adoption can be classified into a binary "yes or no" type of decision. Yet other innovations like contracting were dismissed because they mainly function as tools for financing innovations and cannot diffuse without them. Moreover, cogeneration units do not play a role in our considered scenario 3, which is why they were neither regarded. After all these considerations, five innovations of the initial long list remained, namely 1) photovoltaic (PV) systems with energy storage, 2) smart meters, 3) dynamic electricity tariffs, 4) heat pumps, and 5) electric mobility with charging infrastructure. *Step four* describes the identification of relevant actors, such as households, companies and governmental actors, based on the adjusted innovation list. In *step five*, we determined adopters by assigning all actors to innovations for which they could theoretically appear as adopters. By identifying examples of applications of innovations, the group of change agents was worked out in *step six*. In the concluding *step seven*, all of the above-mentioned results were combined to Table 3 which shows the selected innovations, the relevant actors and the roles the actors play in regard to this innovation: adopter (A), change agent (C), both (A / C) or framework condition (F). Letters in brackets show a weak connection between actor and innovation.

Table 3: Relevant Innovations and Actors

No.	Innovation	House-holds	Munici-palities	Housing Sector	Energy Compa-nies	Distribution System Operators	Com-panies
1.	PV with Storage	A	A / C	[A] / C	C		A / C
2.	Smart Meter	A	A / C	A / [C]	C	C	A / C
3.	Dynamic Electricity Tariffs	A	A / C		C	F	A / C
4.	Heat Pumps	A	A / C	A / C	C		A / C
5.	Electric Mobility & Charging Infrastructure	A	A / C	A / C	A / C	F	A / C

Households are important adopters for all five innovations. *Municipalities* play a dichotomous role. On the one hand, they can adopt all the innovations, for example through installing heat pumps in public buildings or by investing in self-used electric mobility. On the other hand, they play a role as change agents which foster the diffusion of innovation (e.g., through municipal energy consulting). The energy consultants explain different technologies like heat pumps to interested citizens and support their innovation-decision process. The *housing sector*, which consists of companies and households that manage, broker, administer or trade real estate [73] invests into heat pumps, smart metering, PV with storage as well as electric mobility and is, therefore, an adopter. Especially construction firms within the housing sector are advising builders on different heat systems or on how to generate one's own energy. Housing companies also act as change agents, for instance for electric mobility, if they include charging stations in their building projects. *Energy companies* are characterized as typical change agents as they offer innovations or support customers with information regarding installation and operation.

Customers contact their energy company as the first point of contact for all energy questions. The *distribution system operator* plays a central role in the smart meter rollout. It is in the responsibility of each distribution system operator to replace old electric meters by installing smart meters. In the case of dynamic electricity tariffs and electric mobility, the distribution system operator sets framework conditions through the exchange of information regarding grid load or capacity. *Companies* play a twofold role. On the one hand, companies act as adopters, if they demand or implement one of the innovations. On the other hand, they can also become active as change agents, if they supply the innovations, for example by producing (i.e., as the manufacturer), offering (i.e., as a retailer) or implementing (i.e., as an installer) one of these innovations.

The results achieved show that a large number of actors are relevant for the selected innovations. The taken roles vary strongly so that the same actor can be both an adopter and a change agent. One aim of this section was to show the different actors and their respective roles in the diffusion process and to determine the further progress of the research. Instead of lumping together different groups of adopters (e.g., companies and private households), the decision was made to focus on households as one definable group of adopters, as they reflect the only group of actors without an additional role. Therefore, households can be considered for a diffusion study. The focus of the overall project on household applications and individual user behavior supports this focus. The results also necessitate that the other identified actors must be examined since they exert influence on households through their role as change agents.

7.3 Research Design

As our goal with this research is to elaborate on diffusion of innovation theory as a potent general theory, which we adapt to the specific circumstances of the five chosen innovations, we opted for a theory-elaborating qualitative research design [74]. This qualitative approach complements the other, mostly quantitative, approaches shown in the following sections. To this aim, we conducted semi-structured interviews with (potential) adopters (e.g., [75]), as well as expert interviews (e.g., [76]), with change agents such as manufacturers, installers, energy advisors, actors within the housing sector, energy suppliers and distribution system operators. Interviews are particularly suitable means for identifying explanations of key events (i.e., the "whys" and "hows"), as well as the insights reflecting participants' relativist perspectives [77]. Interviews are targeted and insightful instruments. They are targeted insofar as they focus directly on the interviewer's topic of interests and insightful as they identify explanations as well as personal views (e.g., perceptions, attitudes, and meanings). Data triangulation by corroborating interview data with information from other sources is a reasonable approach to mitigate weaknesses and support the validity of the results [77] [78] and was therefore undertaken.

To gain a more differentiated view on how the energy transition affects the different actors and conversely on how the different actors influence the execution of the energy transition, we were regarding the adopters of innovation and its change agents simultaneously. The selection of interview partners considered different aspects. Moreover, different change agents were part of the sampling because they play key roles in the adoption of innovations. Change agents possess valuable knowledge through their individual experiences, as well as their cumulative information about the adopters. Additionally, they can be identified quite certainly compared to opinion leaders who are part of the group of adopters. Regarding the change agents, the selection of interview partners especially targeted those actors, that are in direct contact to the adopters, because they seemed to be capable of providing information about for example the adopter's behavior and motives. Since the opinion leaders are more difficult to identify, this group was not explicitly targeted to be part of the study. Still, some of the interviewed adopters might also play a role as opinion leaders. In total, we conducted 46 semi-structured interviews, 20 of those with adopters, non-adopters, potential and former adopters of the selected innovations and 26 with change agents in different positions (e.g., installers or business representatives of companies within the housing sector). Some of them were identified through the interviews with adopters. The interviews included the different aspects of the four main elements after Rogers [65] and especially focused on the individual adoption processes and the role of different change agents within those processes. The interviews lasted 62:58 hours in total. We transcribed all of the interviews verbatim, leading to 1,054 pages of documents. Subsequently, we conducted various rounds of coding. Most importantly for the purposes of this article, we used codes derived from the diffusion of innovation theory to obtain an in-depth understanding of each of the selected innovations. Often, we complemented this information with additional insights obtained from extant studies on these innovations [79]. Thus, we created diffusion studies for each one of the selected innovations as our main results.

7.4 Diffusion Profiles

This section lays out the diffusion studies for each of the selected innovations. For each of the innovations, we start with a brief introduction to the technology, after that follows the description of its most important attributes (i.e., its relative advantage, trialability, compatibility, observability, and complexity), followed next by the respective social system in which diffusion occurs, then by the communication channels, and lastly by an overall evaluation of the diffusion over time. We present the innovation PV with energy storage (Section 7.4.1) in particularly much detail. The other innovations smart meter (7.4.2), dynamic electric tariffs (7.4.3), heat pumps (7.4.4) and electric mobility and charging infrastructure (7.4.5) were also examined in detail, but due to writing space restrictions, the results are depicted more concisely.

7.4.1 Photovoltaic with Energy Storage

During the last years, photovoltaic systems, which consist of modules and inverters, have become an important part of the electricity supply in Germany [80]. For private households, the performance of photovoltaic systems lies in the range between 3 and 10kWp [81]. Larger systems are usually not suitable for the sizes of rooftops of private houses. Moreover, they come with higher levels of complexity regarding the existing law [81]. In order to integrate volatile supplies of renewable energies into the electricity grids, decentralized energy storage solutions can be considered as solutions that can be implemented within a relatively short period through different solutions: mobile and immobile battery storage systems, heating of drinkable water, process heat and changes in room temperature [82]. This study focusses on battery storage systems, especially the immobile ones (for mobile storage systems, see Section 7.4.5). Since the year 2012, the average electricity price exceeds the feed-in compensation of smaller photovoltaic systems [83]. This development makes it more attractive for adopters to combine their photovoltaic systems with storage solutions in order to increase the own consumption of the produced electricity. Still, the diffusion of photovoltaic systems and storage solutions might differ. Even though this study mainly regarded these products combined, specifics of the individual products are given when necessary.

Both economic and ecological considerations of the adopters can be cited as relative advantages. The majority of the statements here went in the direction that respondents want to achieve some kind of autarky of energy supply. This aim was connected to different considerations, economic as well as ecological ones. Those, who aim to do something good for nature, also wanted the adoption to be economically reasonable. So mostly, the perceived relative economic advantage appears to function as a necessary condition while the ecological perspective appeared to be an attractive add-on in some cases or even the starting point for the general interest into the topic. Some adopters made their decisions simply because of economic reasons, while no ecological reasons were given as the main motivation. Others did not believe that these systems generally pay off, but bought it anyway. The economic reasons are many and varied. The decoupling from the development of electricity prices, for which the adopters expect a further increase in the upcoming years, was frequently mentioned. Raises in electricity prices sometimes appear to be a starting point for thinking about a PV system. A study by Gährs et al. [84] underlines this result, as almost 50% of the 500 interviewed PV owners ranked 'independence from the energy supplier' as a very important reason and the interviewed change agents also confirm the high relevance of this argument. Adopters try to be self-sufficient in energy. Electricity costs can thus be reduced and there is no need to change between energy suppliers. The low-interest level is mentioned by the adopters as a further reason for the acquisition since on the one hand loans are cheap to obtain and on the other hand existing bank deposits are no longer getting interests rates that are perceived to be high enough. Therefore, new opportunities for financial investments are favored and an investment into a PV system with storage seems more profitable to many adopters or was preferred over alternative investments. Using the photovoltaic system as an investment opportunity is mostly the case when it is retrofitted to an existing building. For new buildings, budget restrictions regarding the building overall appear to delay or prevent the installation. In addition, the opportunity to use a PV system as an investment before retirement was considered, in order to reduce costs in periods with lower incomes. As a further financial advantage, the tax refund of expenses in the course of the investment was mentioned. An adopter decided in favor of a system in which, in addition to the storage unit, he also received a free budget of kilowatt hours, which he uses to operate his heat pump, especially in winter. The manufacturers also cite the possibility of using the manufacturer's storage system to achieve an electricity flat rate as one reason why adopters opt for a house storage system. However, the economic benefits need to be considered connected to the initial costs, which in some cases were perceived too high, in others lower than expected. Gährs et al. [84] support this argument, as the high investment was by far the highest ranked reason against a storage system in their study in 2014. According to the interviews, potential adopters principally liked the idea but appeared to be waiting for the technological solutions regarding the photovoltaic systems or the storage system to become more efficient, inexpensive or established. Generally, the view of PV systems as a good investment appears to be less relevant compared to former adoptions in the early 2000s. Today, other motives like the wish for autarky matter, too.

Adopters cite conscious support for the energy transition as ecological reasons or more generally speak of support for the environment. An adopter who lives close to an open pit lignite mine would like to set a conscious example against the generation of electricity from lignite. Another adopter says that it gives him a good feeling to have produced his own electricity and that he is happy about it because it reduces the amount of electricity he gets from the grid. Most adopters were aware of ecological issues regarding the production of the photovoltaic system, which in some cases appeared to reduce their ecological motivation. Green electricity tariffs were in some cases compared to an own system and regarded as an easier, less risky and probably more efficient way to use renewable energies. Still, ecological motives in some cases led to the acceptance of less economic solutions. Overall, the wish for a feeling of independence often positively influenced or triggered the adoption process.

It is difficult to define the trialability of a PV system with storage. Adopters can, however, use their mobile phones with corresponding applications to show other interested parties the control and readout of their system. These mobile phone applications of remote selection offer a first starting point for potential adopters to get in touch with the overall system and gain their first experiences. However, one of the main insecurities regarding the adoption process is the durability of the technological system for time horizons of at least 10 years. This barrier cannot be overcome by trialability. Even adopters, who already implemented photovoltaic systems and therefore trust in one of the technologies, still hesitate to supplement their system with a storage solution. Therefore, the willingness to pay in some cases was lower than current price levels. In the past, the risk of the investment through an uncertain life or usage period was perceived as a reason against storage systems [84]. According to change agents, these concerns are reasonable. Especially with the addition of a storage solution, the payback period is extended and will most likely include more necessary repairs and exchanges of less durable parts (e.g., the inverter), extending the payback period again. Uncertainties among change agents reduce their motivation to influence potential adopters. Some adopters mentioned that they would appreciate some sort of guarantee, at least for the storage system, to reduce their uncertainty regarding the durability.

In terms of compatibility, the structural context, as well as one's own living conditions and lifestyles, must be considered. Since this study focuses on photovoltaic systems for consumers (<10kWp), the grid connection of the individual system usually is unproblematic and therefore not further discussed at this point. From the point of view of the structural context, there are arguments put forward by the adopters who have decided in favor of a system that can be extended later, when the prices for storage facilities have fallen. An adopter already had empty pipes moved from the roof into the cellar during the construction of the house ten years ago, so that the complete system including storage can be retrofitted later. Another adopter is planning to expand his PV system if a side roof is renovated, as work is already being carried out on the roof in this connection, thus creating synergies. However, generally, the adopters need buildings with suitable roofs. Tenants showed no motivation to invest higher amounts of money into systems for buildings they rent. For new buildings, the direction of the roof can be planned according to optimal placement of a photovoltaic system, but with existing buildings, this is not possible. In addition, there is a need for space in the utility rooms, which are usually planned to be quite small. In some cases, monument protection prevents the adoption of the photovoltaic system. Compared to other innovations like heat pumps, the retrofit of a photovoltaic system appears to be relatively simple to implement. However, the installation comes along with visual changes out- and inside the building, which is not always appreciated by potential adopters. Concerns about the security of the system were barely raised by the interviewed adopters, in line with prior findings [84]. All this leads to strong compatibility of the system with the structural context. The idea to combine a photovoltaic system with, for example, a heat pump or a private charging point for an electric car supports the relevance of compatibility. Even though such combinations are not always in place, adopters and change agents appreciated the idea of connecting different applications within the household. In order to achieve higher levels of compatibility between the photovoltaic system and the energy storage solution, simultaneous adoption is advantageous. Later additions of energy storages to existing photovoltaic systems were perceived to be less fortunate because then some additional components are needed. Two types of adopters can be considered from the point of view of the compatibility of one's own living conditions with the newly acquired system. The first group is not able or willing to change their behavior. For them, the implementation of a photovoltaic system does not have any impact on their own living conditions. They are willing to accept higher costs, for example through larger and more expensive storage. The other group of adopters tries to change their lifestyle in order to achieve high compatibility with the used PV system with or without storage. This is done by shifting consumption such as the utilization of washing machines and dishwashers in times when a lot of electricity is produced or by purchasing particularly energy-efficient appliances. Through these manual adjustments costs for automation and storage systems can be avoided and therefore make the photovoltaic system itself more profitable. In addition, the purchase of an electric car to use the self-generated electricity for mobility is implemented or considered by the adopters.

The observability of this innovation plays an important role in the innovation-decision-process at various points in time. In the run-up to the decision, the increased visibility of PV systems on rooftops or as ground-mounted systems can influence potential adopters, currently leading them to conclude that this technology has been tested and recognized, even if some do not like the look of the panels. Many adopters have reported that they have looked at systems from relatives and friends or have inspected several of the installer's built-in systems. In only a few cases, friends and neighbors of the interviewed adopters show an interest and request further information after the implementation of such a system. An adopter reported that his brother-in-law had explored the PV system with him and had then installed such a system as well. The first adopters influence further diffusion with their adoption decision. In addition, the possibility of controlling the PV system by mobile phone or with a computer, possibly in combination with an energy management system, triggered the interest of many respondents and demonstrates the functionality of the system. Even though these applications are isolated from other applications regarding the housing context, the adopters appear to keep checking them frequently.

Adopters assessed the complexity of innovation differently. It was striking that a large share of the actual adopters had a technical background and that they usually had no problems with the technical complexity or considered it a challenge to deal with the matter in detail. Still, even among these adopters, there were insecurities regarding the judgment of different storage system solutions. An adopter specifically addressed the complexity of decision-making, thereby highlighting a need for independent advice and assessment of offers. Bureaucratic and tax procedures were described as complex to many adopters. In particular, the tax treatment was usually passed on to a tax consultant. However, this results in higher costs. For some adopters, the registration of the PV system with storage at the responsible distribution system operator was complex. Others benefited from the preliminary work of their installers, who had prepared the necessary documents for them so that the complexity for the adopter was strongly reduced. Installers also facilitated the combination of photovoltaic system and storage solution by offering these products in bundles.

The social system is the second main analysis unit and plays a major role in the process of adoption. Within this element, the results of the study can be separated into insights about the adopters and information regarding the role of different change agents.

In addition to the already stated information about the adopters' perceptions of the innovation attributes, also revealing details about the adopters' motives and contexts, some general information about the interviewed adopters can be stated. The adopters came from very different contexts. Generally, the energy transition appeared to be a familiar topic and therefore PV systems were known. While many adopters were able to use private contacts to gain further information, others found themselves in difficult environments. An adopter from a lignite mining area reports that some understand the PV plant as a statement against lignite and one is stamped as an opponent. Technical background of the adopters turned out to be helpful for the adoption decision, as there was a basic understanding of technology and an interest in dealing with specific characteristics of the system. Some adopters also stated to be familiar with these technologies since they have work-related knowledge regarding renewable energies or know photovoltaic systems from other areas of application, like for example for a camping van.

The role of the change agent is embodied through different actors from different companies and branches. The following change agents seemed to play an important role within the decision processes of the adopters:

- Installers of photovoltaic systems
- Producers of photovoltaic systems
- Politics and funding bodies, especially the German Kreditanstalt für Wiederaufbau (KfW)
- Energy advisors
- Different actors within the housing sector such as architects and building companies
- Energy providers

In the decision process of an adopter, the personal conversation and the confidence-building were important for an installer. It was mostly spoken with local companies since these are established, other known customers have already worked with them and in the case of a breakdown, the closeness was perceived as an advantage. An adopter suspects that smaller installers have no opportunity to train their staff. Much is mastered in learning-by-doing, so mistakes are more likely. An adopter complained that the quality of the installers was poor and that he could not inspect them in advance. Generally, installers still active on the market experienced a lot of concurrence through less professional installers in times of high demands caused by high subsidies. The installers have important roles in filtering the information regarding the technological systems for the potential adopters, delivering only the necessary details, leaving out technological details to reduce complexity, and tailoring their offers to the adopters' needs and support the decision process.

With regard to the producers, trust issues, especially toward certain foreign producers became obvious from both sides, adopters and other change agents. With regard to guarantees, there is uncertainty as to whether the in some cases long guarantee periods can really be fulfilled in the event of an occurrence, as many companies have already gone bankrupt. Nevertheless, an adopter has purchased additional guarantees in order to be covered over the entire period of use. Other adopters have made their own economic calculations and tried to take wear into account.

As the main funding body, many adopters mentioned the KfW. Even though this actor appears to be well known in the German institutional context where data were collected, adopters mentioned problems to receive the offered funding. One adopter was disappointed that he could not fall back on a KfW promotional loan for his PV system, as no bank in his vicinity wanted to process the loan application due to the low level of financing and the corresponding high effort. As a result, the adopter was unable to obtain repayment subsidies. Because of the low-interest levels, banks offered loans for the same prices as the KfW conditions. In order to avoid bureaucratic efforts, these offers appeared more attractive. KfW speaks of its market incentive program "Renewable Energy Storage Systems", which was launched in May 2013 and ended on schedule at the end of 2018, as a prime example of a successful market-based promotional program, since it has now been possible to achieve a self-sustaining market, established technical standards and lower unit prices [85].

Direct subsidies for the initial investment, like subsidized loans and grants to the costs of purchase, were barely used by the interviewed adopters. The necessary effort for the application and possible regulations mostly scared off the adopters. In addition, the adopters could finance their installation usually themselves or paid these from their savings. In one case, a regional support program was not used, as the framework conditions with a low level of own consumption did not correspond to the ideas of the adopter. A local initiative with funding opportunities support the installation of photovoltaic systems through consulting and advice but use their financial support for other energy efficiency related measures. Subsidies like feed-in tariffs or financial support for the own consumption of the produced electricity were used frequently. One adopter addressed the feed-in tariff during the interview, which is guaranteed over the term, but in his opinion, there is still uncertainty as to whether this tariff will not be reduced in the future.

Energy advisors provide anyone interested with advice regarding a broad set of areas, also including electricity consumption, heating, ventilation or insulation. Looking at new, as well as existing houses, energy advisers often found more efficient ways to reduce expenses for different energy carriers at lower costs. Institutions offering consulting and subsidies have to act independently and therefore only give advice but do not directly recommend products or companies. Other energy advisors also include the selling of energy systems into their range of services, as well as construction, accompanying services, aftercare operations and the support of the adopters in their interaction with different change agents (e.g., building companies or installers). Moreover, they support the adopters with information on the sizing and help with the connection of different energy-related systems (e.g., heat pumps or charging stations for electric cars). Their possible influence highly depends on the timing of their assignment and when they are included in the planning and decision processes.

Concerning new buildings, architects and companies offering prefabricated houses seemed to be important contacts in the planning process of the building and therefore support the potential adopters with many necessary decisions. However, photovoltaic is only one of many topics that can come up within this process. So even if these actors seemed to be capable in advising their customers or coordinating appointments with installers, their focus lies on the building as a whole and therefore does not necessarily include too much effort into photovoltaic. Internal arguments or issues prevented these change agents from putting further effort into this topic. For example, a company building prefabricates houses nationwide does not tend to cooperate intensively with local installers and cannot compete with them regarding their prices. So, even if they offer the photovoltaic systems themselves including taking care of all organizational issues regarding the installation, they are often asked to only include the infrastructure for a photovoltaic system, leaving the installation to the home-builder. This option is appreciated by the customers since a lot them are households with limited budgets and therefore focus on more elementary aspects in the building process first, leaving the photovoltaic system as a possible addition for a later point in time, even if it is useful and more convenient to include it all at once into the building process. Architects also consult customers who wish to refurbish existing buildings. However, similar to the energy advisors' suggestions, they have a broader view of the building, which does not necessarily prioritize PV system. Moreover, not all architects are certified to build specific subsidized houses and therefore need to bring in another architect, which was perceived as an additional bureaucratic hassle by the adopters.

Adopters use a variety of communication channels to obtain information. It is therefore studied as the third analysis unit. For many adopters, the innovation, in general, was already known quite well. The specific research on the internet began in order to gain a rough overview of manufacturers, products and installers. In addition to telephone calls to the manufacturers, brochures were also ordered to obtain further information. Important contact points for potential adopters are the expert forums on the internet, where actors share their knowledge and experience about implementation. One expert forum even provided an opportunity to review the offers for the adopter. These forums appear to be a good opportunity to verify information from producers' websites in case of existing trust issues regarding their information. Visits to customer installations and discussions with customers about the installers are important information channels. However, a central part of the communication process is the personal discussion with the installer. Many adopters report that trust in an installer is the most important criterion, but is also the most difficult to identify. The installers who appeared to give advice independently and did not appear to focus simply on selling their products were positively emphasized.

Looking at the diffusion of photovoltaic systems as the fourth unit of analysis, there were approximately 1.7 million photovoltaic systems in Germany overall in 2018, 162 thousand of them in Lower Saxony. 6,161 of the systems in Lower Saxony have been installed in that year, almost all of them on roofs. After growth rates drastically went down in the years before 2015, they slowly have been increasing again since then [86][1]. Meanwhile, the product itself developed further, becoming more efficient and more cost-effective [87], thereby increasing the relative advantage, as well as more flexible regarding the use in less optimal construction conditions. In addition, smaller solutions, e.g., for balconies, which easily can be plugged into a regular socket, were invented. These developments have positive effects on compatibility.

The diffusion of photovoltaic systems in the past has been shaped through political interventions, specifically, the feed-in tariffs regulated by the German Renewable Energies Act[2] [88] or minimum requirements pertaining energy-related decisions for new buildings regulated by the German Energy Saving Ordinance[3]. Lower feed-in tariffs reduce the perceived relative advantage, while regulations for new buildings generally support the diffusion positively, but do not concern the existing building stock. Another important factor influencing the diffusion of photovoltaic systems are the electricity prices, also being linked to political decisions [83].

The share of photovoltaic systems installed jointly with energy storage solutions experienced a sudden increase in Lower Saxony only a few years ago. While in the year 2014 only 16% of new photovoltaic (<30kWp) systems were combined with energy storage [89], in 2015 the share was already 53% [90]. In the following years, 2016 and 2017 that share seemed to level off at 45-46% [91] [83]. The interviews revealed that the adopters were aware of sinking prices of storage systems within the last years and therefore perceiving the storage solutions as more attractive. According to the change agents interviewed, market diffusion is only just beginning. In the long term, the average expectation is that more than half of the single-family and multi-family houses in Germany will have a PV home storage system.

However, storage systems are not only used in combination with new photovoltaic systems but also represent a possible extension for existing systems. From 2013 until 2017, 80-90% of storage solutions were installed with new photovoltaic systems and 10-20% with existing systems [83]. These systems are probably the ones that were built after 2009 since their production is not completely used for feed-in but already include a share of own consumption. These shares might change after 2021 when the photovoltaic systems installed in 2000 run out of their feed-in compensation which is fixed for 20 years (EEG § 22) [81]. It is expected, that these photovoltaic systems will still be productive or that their productivity can be restored by assembling new parts [81]. This additional installation of storage solutions for photovoltaic systems losing their feed-in compensation will probably

---

[1] Summarized and processed data from the Agentur für Erneuerbare Energien (www.foederal-erneuerbar.de) who work with data from the Bundesnetzagentur für Elektrizität, Gas, Telekommunikation, Post und Eisenbahnen.
[2] Erneuerbare-Energien-Gesetz (EEG)
[3] Energieeinsparverordnung (EnEV)

go on until the year 2035 when the subsidies of the systems from 2015 expire when there already were significant shares of energy storages. The reason for this is the introduction and reduction of feed-in tariffs. With the introduction of the EEG in 2000, high feed-in tariffs were set for PV systems, which were intended to stimulate investment in the technology and drive expansion. In the following years, the manufacturing costs for PV modules fell sharply, so that the subsidy rates in the EEG were adjusted. The reductions resulted in subsidy rates falling from 50.7 cents/kWh in 2000 to 11.11 cents/kWh today [92]. With the strong reduction below the level of electricity purchase costs, an incentive for own consumption emerged, so that more PV systems with storage facilities are being added. Another possible application for home storage systems goes beyond the optimization of own consumption and aims at the use of electricity price differences within one day through dynamic electricity tariffs. A storage facility could also be of interest without a PV system. However, this is only feasible if storage prices fall sharply and unlikely to be attractive for private households in general. The change agents interviewed see the potential for these applications very differently. While energy suppliers and manufacturers see a large market and new business models, the potential of the interviewed installers is estimated to be much lower, as the limited capacity of the storage systems does not lead to a high economic benefit.

To summarize, the pursuit of autarky of energy supply represents the main motivation of the adopters. Ecological aspects play a complementary role behind the more important economic aspects. The high costs and concerns about efficiency are the main factors that prevent adoption. A large number of different actors play an important role in diffusion. The personal discussion with these actors is central for the adopters in order to build trust. Continuing price degression, technological optimizations, and increasing public awareness are leading to the removal of barriers and an increase in the diffusion of storage systems. The experts interviewed assume that decentralized storage systems will be an integral part of single-family homes and apartment buildings in the future. An investment in a home storage system will, therefore, be of interest to a large number of households over the next 10 years. The trend toward sector coupling, in particular, the coupling between PV storage systems and electric mobility, suggests that synergy effects can be expected.

7.4.2 Smart Meter

A smart meter is a digital counting and measuring technology, which is to successively replace the conventional electricity meters. It consists of a measuring device and a communication module, the gateway [93]. In contrast, a digital meter does not have a gateway and therefore cannot send data or receive commands. Smart meters are electricity (or gas) meters that offer further functionalities beyond the mere measurement of consumption, e.g., they make it possible to display consumption behavior over the course of a day, week or year. They can provide end consumers, distribution system operators and producers with the necessary consumption information, serve to transmit information for smart grids and create appropriate incentives for end consumers to improve energy efficiency ([94]; [95]). Smart meters enable the introduction of dynamic electricity tariffs. As an interface to the consumer's electric supply, they offer automation potential for controlling shiftable loads like electric storage heaters, heat pumps, and electric vehicles as well as decentralized generators like photovoltaic systems [96].

A smart meter offers a relative advantage as it provides feedback for the customer about the current consumption and therefore the opportunity to save energy and money by behavioral adjustments. The aspect of compatibility is therefore particularly important for the devices that interact with the smart meter, e.g., controllable appliances. With respect to complexity, it is important that the smart meter's control system be designed to be easily accessible to the consumer. Otherwise, a complicated control system may frustrate users and lead them to abandon the device. The trialability of smart meters can lead to the removal of barriers to their use as it results in reduced uncertainties [65]. In the case of smart meters, it is difficult to imagine that they could be tried out, as these devices are usually not visible or cannot be installed on a trial basis. For the same reasons, observability is also not given.

Adopters of smart meters are, according to the Metering Point Operating Law[4], those parties with high consumption or those who deal with the generation of their own renewable energies. These can be households, municipalities, the housing sector or companies. The distribution of smart meters is the responsibility of the distribution system operators who therefore act as the central change agent of this innovation. They receive support from smart meter manufacturers, installers, energy suppliers and local authorities who can advertise the installation, offer advice and in doing so remove barriers.

No special communication channels are recognizable since it is to be expected that the communication will originate from the responsible distribution system operator. The diffusion of smart meter devices is legally integrated into the Act on the Digitization of the Energy Transition[5] of August 29, 2016. With this act, the Federal Government pursues the goal of linking generation and consumption more closely in the electricity grids. The law provides for a gradual introduction (so-called rollout) of smart meters, starting with large-scale consumers of electricity (> 10,000kWh per year) and larger systems subsidized under the EEG (7 to 100kW of installed capacity) in 2017. By 2021, the mandatory installation is to be extended to include consumers

---

[4] Messstellenbetriebsgesetz (MsbG)
[5] Gesetz zur Digitalisierung der Energiewende

with at least 6,000kWh per year [97]. The vast majority of households will not reach this limit. Only an average five-person household that lives in a single-family house and has an electric water heating will usually reach a value of 6,300kWh. A comparable household without electric water heating (5,000kWh) is well below this limit. If the household consists of five persons in an apartment building, the limit is exceeded neither with electric water heating (5,600kWh) nor without (3,600kWh) [98]. Since the distribution system operator installs the smart meter, the respective customer has no choice regarding the type of device. However, certification of three different smart meter gateways is a prerequisite for the start of the rollout. At present (April 2019), the German Federal Office for Information Security has successfully certified only one manufacturer [99]. It can, therefore, be assumed that there will be delays in the rollout and that the actual plan cannot be fulfilled.

The distribution system operators expressed their opinion with regard to smart meters that the legal obligations would be fulfilled. However, not every household will receive such a meter, as in many cases a digital meter is sufficient. The digital meters are already being installed.

To conclude, smart meters for large consumers can provide an incentive to save electricity by showing the current consumption of the devices used. It is important that the devices are compatible with the smart meter so that communication and control are possible. For many households, it will not be necessary to install a smart meter, as digital meters will be legally sufficient.

7.4.3 Dynamic Electricity Tariffs

Dynamic electricity tariffs are supposed to play an important role regarding the management of electricity demand by integrating volatile renewable sources and maintaining stability within the grid. The plan is to create financial incentives for favored user behavior through electricity tariffs. This strategy is already in place for companies exceeding a certain electricity consumption. The legal basis for these tariffs in Germany is § 40 paragraph 3 of the Energy Economy Law[6] requiring electricity suppliers to offer such tariffs, and incentivizing a reduction or the management of energy consumption. There is a variety of different designs for these tariffs: time-of-use pricing, critical peak pricing, real-time pricing, direct load control, emergency demand response, curtailable load, interruptible load, demand bidding, and usage-bound tariffs. These different tariffs work with varied instruments, different pricing systems or other financial incentives [100] and create diverse reactions or consequences for the users. We focus on those tariffs that potentially motivate users to postpone their individual electricity use to different points in time.

---

[6] Energiewirtschaftsgesetz (EnWG)

For example, usage-bound tariffs might lead to a reduction of energy usage overall but have no noticeable potential to postpone energy consumption since the periods considered for pricing differ from a month to a year. A more detailed contemplation on the individual load shifting follows in Section 9.1.1.

As mentioned above, the benefit for users is possible financial incentives. However, any actual savings depend on the tariff's design and on the equipment within the household. A thermal storage system or bigger electrical loads such as heat pumps or electric cars in combination with a smart measurement system lead to a higher potential for load shifting and therefore the realization of financial benefits [101]. It also needs to be considered how the reaction to price signals takes place; automatically, partially automatically or manually [102]. A manual reaction leads to a loss of convenience through the need for monitoring and pressure to change behavior, while automatic reactions are only facilitated by more expensive technical equipment [101]. Moreover, some interviews revealed that the adopters are very critical of higher levels of automation within their houses. Generally, the anticipated advantage depends on the expectation regarding the development of electricity prices. For adopters expecting increasing price levels, the dynamic tariffs appear more disadvantageous [103]. Regarding the compatibility, there are essential restrictions. For some tariffs that only include two different price periods and can hardly be called dynamic the implementation of a so-called multi-rate meter is enough. For tariffs with higher levels of flexibility, there needs to be an intelligent metering system (see the previous section on smart meters). The perceived complexity probably depends on the tariff's design. A trialability does not seem to be given without entering into a contract. The visibility is provided through the advertisement of electricity providers offering such tariffs, but it currently does not seem to play an important role in the adopter's social environment or conversations since it is not a present topic for the interviewed adopters themselves.

The electricity providers offering these tariffs are important change agents. Some of them use online tools to simplify the search and information process of potential adopters (e.g., [104]). Problematic, regarding the change agents' motivation to offer such tariffs might be, that it is the grid operators who must deal with problems caused by the integration of volatile renewable energies and not the electricity providers directly unless grid operator and electricity provider are combined within one company. Moreover, electricity providers need to go through considerable changes in their internal processes, like changing their whole balancing procedures, in order to offer dynamic electricity tariffs [101]. Electricity providers report that they offer or have offered tariffs with two different price periods (high and low tariff), but that demand is very weak or so low that the product has been discontinued. Some electricity providers, therefore, do not believe in dynamic electricity tariffs because they doubt that the majority of customers have an interest in dealing with electricity prices or are willing to shift consumption. Two electricity providers reported that they would need price signals from an upstream grid level for such an offer and thus become dependent on it, which is not in their interest. The electricity providers that can imagine a commitment in the area of dynamic electricity tariffs currently see few application possibilities. They attribute this to the lack of infrastructure, as smart meters are not available until now. Some electricity providers, who are also distribution system operators, see the combination of smart meters, dynamic electricity tariffs, storage, and electric mobility as a way of intelligently shifting loads in the grid and thus enabling stable grid operation. Another change agent turned out to be within the building sector. For example, companies offering prefabricated houses seem to play a role in the adoption process, by directly connecting the usage of, for instance, a heat pump to the offer of a tariff with two pricing periods and including the necessary infrastructure (multi-rate meter) into the construction schedule. That way the adopter can pay less for the energy the heat pump uses at night.

Regarding the used communication channels, according to the interview partners, there was little conversation about this topic to others. Adopters or those who did not adopt became aware of these tariffs directly through the named change agents.

So far, the diffusion of dynamic electricity tariffs has not gone very far. Most parts of the population have never heard of them or barely know anything about their advantages or disadvantages [105]. Even though there appears to be a receptive attitude, opinions regarding the personal use of such tariffs vary. There are doubts regarding the possibility to adjust their own behavior or rejections to do so because the expected potential for savings is not perceived to be high enough [105]. The study by Gerpott and Paukert [103] showed that higher levels of flexibility like shorter notice times or higher differences between the highest and lowest prices appeared to be less desired by potential adopters. Other results did not find that this difference is important but still confirmed that users prefer the current tariffs over dynamic ones and if the tariff needs to be dynamic, they prefer the less dynamic ones [106]. Looking at the change agents, especially the electricity providers, most of them offer the already named tariffs with two pricing periods. These tariffs already started to diffuse decades ago along with night storage heaters and only require multi-rate meters, not necessarily smart meters. Even though more designs that are dynamic have been tested in model regions [107], there is no significant diffusion so far. Only 3% of the suppliers offered dynamic tariffs reflecting the spot market prices in intervals to private households [108]. The 268 electricity providers offering tariffs in Lower Saxony in 2017 are overall offering less than ten dynamic tariffs beyond the ones with only two pricing periods. In some cases, these tariffs directly pass the market prices on to the users but do not announce the upcoming price changes.

Thus, these tariffs do not give the users a chance to adjust their behavior accordingly (e.g., [109]).

Even though dynamic electricity tariffs potentially play a key role within the energy transition by providing incentives for load shifting and therefore supporting the integration of volatile renewable energy sources, they are barely known and have hardly diffused among private households. Currently, the expected financial advantages are not perceived to be high enough compared to the trouble of implementation, for both, change agents and adopters.

7.4.4 Heat Pumps

Heat pumps are heating systems that mainly use the energy of the surrounding environment. For the pump to work, there is only a minor need for electricity or gas. A heat pump's functional principle is based on the so-called Joule-Thomson-effect, a physical effect that drains thermal energy outside the building and gives it off within the building. Possible thermal energy sources are the soil, the air outside or the exhaust air from the building. The most common heat pumps use energy from these thermal energy sources, which is drained by using refrigerant on a low-temperature level and then heating it up through mechanical or thermal compression until it is suitable for heating purposes such as space and water heating. Some also facilitate air conditioning [70]. Different studies present heat pumps as a key technology for a transition of the heating system away from fossil energy sources [70] [110]. Even though the focus of NEDS is on the electric energy system, heat pumps were chosen to be considered as well because this technology contributes to an integrated energy system based on electricity [111].

The relative advantages perceived by potential adopters are diverse, comprising expectancy of little maintenance, effectiveness of the technology, independence from fossil fuels such as gas or oil, resource conservation, and sustainability. The latter aspect appears to have a lower priority to most households. Moreover, the level of sustainability depends among other factors on the electricity mix the heat pumps use. Change agents clarify that there are not necessarily any financial advantages. The initial costs are not always clear to house builders, because they are included in total prices for the new building and therefore hardly comparable to alternative solutions. Looking at the compatibility of heat pumps there are difficulties on the one hand, especially regarding older buildings. In order to use a heat pump, the building must be well-isolated and equipped with extensive radiators. When building a new house, initial costs for a chimney or a gas connection can be saved. For existing buildings, this advantage does not apply and reduces the competitiveness of the heat pump compared to other solutions once again. Overall, change agents confirmed that currently, heat pumps have no significant relevance for existing buildings. Some developing areas are even planned without a gas connection, which reduces the choices of possible heating systems. In others, the buildings are planned to be proximate to each other, which may lead to insecurities surrounding noise emissions of certain kinds of heat pumps. Apparently, the compatibility regarding the context of building and area matters, yet it can do so in both, a positive (i.e., adoption-fostering) or negative (i.e., adoption-obstructing) manner. The connection to an own PV system, which allows households to use the electricity produced by that system to fuel the heat pump, is valued by the adopters. Change agents have to relativize the adopters' expectations considering the fact, that the PV system's production is not that high in the heating period. Overall, the compatibility heavily depends on the building context. Moreover, legal requirements for new buildings often trigger the adoption, even for those adopters who did not initially plan to install a heat pump. With respect to mental compatibility, the perception of electricity appears to be shifting. In the past, electricity was primarily perceived as a precious and slightly negatively connotated resource, that needs to be saved. Today, consumers view electricity more positively, according to the change agents. Given occasional electricity surpluses and the need for integrating electricity into the heating sector, consumers are more willing to apply electricity-intensive technologies than they used to be. Heat pumps are also more convenient for consumers because there is hardly any need for regulating them on a daily basis. Observability, as well as trialability, might also obstruct a broader diffusion. A building's heating system is usually not obvious and mainly not seen by, for example, visitors of a building. Trying out a heating system would require an extant installation. One aspect where trialability appeared to play a role was by adopters listening to installed heat pumps in order to experience potential noise emissions. The complexity in the purchasing context seems to be reduced by parties such as building enterprises or architects supporting the building process through suggestions, fitting the heating system to building and personal needs and coordination of different artisans. In some cases, the building processes leave little room for individual solutions and personal interventions by the final adopters.

Possible *adopters* of heat pumps are those organizations and individuals who own or manage new and redeveloped houses, company facilities, or even entire districts [112] [113]. Relevant *change agents* for the diffusion of heat pumps comprise production and trading companies specialized on heat pumps, as well as organizations in the building industry, specifically companies offering prefabricated houses, and the architects of new buildings. For these new buildings, legal requirements regarding their energy supply often appear to drive change agents and adopter's commitment to heat pumps [110]. Heat pumps are in some cases seen as a comparably easy way to satisfy these regulations. Another relevant change agent group are the energy suppliers that are active in the field of heat supply. These players operate heat pumps in contracting in single-family and multi-family houses.

The energy supplier, who receives a monthly fee in return for his services, usually handles planning, financing, installation, maintenance, and operation jointly. Instead of high initial investment, many customers find the monthly constant rate in the contracting model more pleasant.

Regarding the relevant communication channels for attention and further information on heat pumps the adopters listed very diverse channels: attention through articles, further information over specific research on the internet and brochures of suppliers. Information found on the internet often appeared to be negative or contradictory. Some adopters tried to verify this through consulting of change agents. Integrated offers by building companies, which included heat pumps, appeared to be decisive for the adoption decision. In some cases, these offers were verified by experts (e.g., a plumber).

The first heat pump in Germany was installed in 1969. Since then this technology has been developed further, and initial technological failures and low performances appear to be overcome [114]. In 2017, heat pumps have been used in more than 42% of new residential buildings in Germany [115]. Some building companies sell heat pumps as a standard solution. However, new buildings only represent a small part of the existing building stock. Most buildings are old and therefore mainly not suitable for the installation of a heat pump without profound renovation [110]. The fraction of buildings with heat pumps within the whole German building stock, including residential and non-residential buildings, is only 2% [110]. For the future *diffusion* of heat pumps, different factors appear to be relevant, mostly the erection of new buildings [70], but also the development of the prices of alternative heating technologies compared to the prices of heat pumps. Other influencing factors are the developments in fuel costs and electricity prices. Because of the dependency of the usefulness of the installation of heat pumps on the building context, the diffusion of heat pumps over all existing buildings is closely linked to the regeneration of the building stock and therefore inherently slow. Even though the numbers of installations for new residential buildings are remarkable, new buildings only constitute very small shares of all buildings.

Overall, heat pumps do not differentiate from other heating systems through the direct results of their performance regarding the room temperature or wellbeing of the inhabitants. The most relevant factors influencing the diffusion of heat pumps are regulations in favor of this heating system and the building context the adoption is considered in. The main challenge for a far-reaching diffusion will be the useful implementation of heat pumps into existing buildings.

7.4.5 Electric Mobility and Charging Infrastructure

Electric cars have been produced since the late 19th century. However, after the invention of the electric starter for combustion engines, their production hardly played a role in terms of numbers [64]. Electric mobility is therefore by no means a new invention, but the technology has been "rediscovered," as it enables mobility without the emission of CO_2 and harmful gases [116] if the electricity used comes from renewable sources. In the smart home, the charging infrastructure can influence the size of the storage facility and the PV system if adopters want to produce the electricity for their electric car themselves. We, therefore, focus on the diffusion of the charging infrastructure, knowing that electric cars and corresponding infrastructure are closely intertwined.

Electric mobility can achieve a relative advantage in terms of operating costs and social prestige compared to existing mobility. In terms of operating costs, electric cars can sometimes be operated cheaper than conventional cars [117], for example, if the batteries are charged via a cheap electric car tariff from the electricity provider. Another possibility is free public charging stations, for example, the ones that some supermarkets offer to their customers. If electricity from regenerative sources is used to charge the electric car, the avoiding of emissions is another relative advantage. A further relative advantage can be achieved in terms of prestige since the drivers of an electric car opt for environmentally friendly mobility and show this by using this car. In this way they can gain recognition for their social environment. The German government and the automobile manufacturers subsidize electric mobility by giving buyers of electric cars a buyer's premium of 4,000 euros and buyers of hybrid cars 3,000 euros. These subsidies are given to speed up the rate of adoption of innovations by increasing the relative advantage of the new idea [65] as these cars are more expensive than conventional cars. Compatibility is a critical aspect since previous behaviors with classical mobility cannot be transferred one-to-one to electric mobility. Electric cars do not reach the ranges of classic cars and the network of charging stations is not yet dense enough to guarantee ubiquity of recharging opportunities along the way. Further challenges are the various charging adapters and billing systems, so that not every user can use every charging station. Standardization, both in terms of charging infrastructure and billing systems, is necessary for the further diffusion of electric mobility. We conjecture that adopters of electric cars will largely install their own private charging station, as it is to be expected that a large part of the charging processes will take place at home. The purchase, installation, and maintenance of these wall boxes entail additional costs. In connection with the currently higher costs for an electric car, the acquisition costs of the entire system represent a high barrier for potential adopters. Residents of multi-party houses without parking at the house are at a disadvantage, as there is no private charging possibility for these cars. In the case of electric mobility, complexity plays a subordinate role, since electric mobility does not present complex difficulties for the user, but strongly follows known patterns of classic car use. Trialability is given in the case of electric mobility since test drives are possible at car retailers or at mobility events. Energy suppliers partly offer the rental of electric vehicles and participate in information events. Car sharing providers are now increasingly offering electric cars in their fleets. These possibilities of trialability have a positive influence on the adoption of electric mobility. The observability of electric mobility is given, since in road traffic the cars with electric drive must have a special license plate with the identifier "E". A further aspect of observability is the increasing construction of charging stations, which, due to their design and exposed location, are noticeable at important public facilities such as railway stations or car parks.

Households are typical adopters for electric mobility as well as municipalities, the housing industry, companies, and energy suppliers if they use electric mobility themselves. Nevertheless, due to their position, they can also be change agents. Municipalities can promote the purchase of an electric car and designate areas for charging infrastructure. The housing sector can include electric mobility in the planning of new construction projects and provide space for these. Companies can promote electric mobility by providing charging infrastructure for their customers and employees, and energy providers can ensure that the charging infrastructure is set up, offer attractive charging rates and get involved in car sharing. Distribution system operators play a special role in the diffusion of electric mobility as they determine framework conditions. Due to their activities in the distribution grid, these players can prohibit the installation of charging stations if this could endanger the stability of the grid. Particularly in the case of many simultaneous charging processes in households, for example in the evening hours, the utilization of the grid infrastructure can get into a critical area. An important future task for the distribution grid operators will thus be to prevent overloading of the electricity grid. Increased grid expansion and intelligent load management offer opportunities. An amendment to the Low Voltage Connection Ordinance[7] will allow the distribution system operator a co-decision right in the construction of charging stations. Charging stations with an output of more than 12kW may only be commissioned with the consent of the grid operator. When new power lines are laid in new development areas, large-scale electric mobility is not considered, as otherwise, the costs would be prohibitively expensive. These limited grid capacities can represent a barrier to the future diffusion of electric mobility.

With respect to relevant communication channels, many energy suppliers report on electric mobility in their customer magazines and advertise their electric mobility days on the internet, with advertisements in newspapers and with flyers. Personal

---

[7] Niederspannungsanschlussordnung (NAV)

conversations are particularly important during the show days. However, this is only an excerpt of possible ways in which adopters can obtain relevant information on electric mobility. For example, before a PV system is installed, a personal meeting with the installer is often held to determine whether an electric car is available or to be purchased. An electric car is a large consumer of electricity in a household and therefore influences the choice of the sizes of both the PV system and the related storage.

In recent years, virtually all automobile manufacturers have added series models in the field of electric mobility to their portfolio. The reasons lie in a mixture of stricter environmental legislation, public subsidies, falling prices, especially for batteries, and at the same time higher achievable efficiency in terms of increased power and greater range, as well as improved charging infrastructure. In addition, customers' willingness to pay tends to grow with environmental awareness [116]. On 1 January 2018, 53,861 purely electric vehicles were registered in Germany. Measured against the total number of approximately 46.5 million registered cars, the current diffusion of purely electric vehicles is 0.12%. However, compared with 2017, the growth rate stands at 58.3%. In Lower Saxony, there were less than five thousand purely electric vehicles per 4.7 million cars, which correspond to a share of 0.10%. The growth rate compared to the previous year is 54.6% [118]. Another characteristic of the diffusion of electric mobility is the expansion of the charging infrastructure. According to BDEW figures, more than 16,100 public and semi-public charging points are available in Germany at the end of 2018, 12% of them as fast loaders. More than three-quarters of the charging stations are operated by the energy industry [119]. The diffusion of electric cars is currently still low but is growing at a high rate. The further, necessary expansion of the charging infrastructure can remove barriers to compatibility and thus contribute to greater diffusion. A higher share of renewable energies will ensure that the sustainability of electric mobility improves and that its diffusion increases.

To summarize, electric mobility can achieve relative advantages in terms of operating costs and emission avoidance. The charging processes, which will mainly take place in private households and here mostly at certain peak times, will place a heavy load on the existing distribution grid. In addition to strengthening the distribution grids, the solution requires above all control of the charging processes at the grid level.

7.5 Discussion

This section begins by summarizing some practical implications of our diffusion studies. Thereafter, a critical reflection follows before we close the section in suggestions for future research.

7.5.1 Practical Implications

Some practically relevant insights mainly concern PV systems with energy storage and heat pumps. Different groups of change agents share and support the opinion, that extensive technology within buildings does not come first when following the goal of energy efficiency. The type of building coverage and attributes like insulation are more effective and therefore prioritized in the recommendation for the adopters. Most single-family houses are not individually designed by architects, but sold by building contractors and companies offering prefabricated houses. Architects come into play when the building owners have special preferences. Different building processes depending on the involved change agents portray different levels of transparency. Moreover, the opportunity and encouragement to influence given standards differ significantly. Therefore, the adopters do not necessarily participate in all kinds of decisions during the planning and building process proactively. This also includes decisions concerning energy-related products. PV systems in some cases appear to be a driver or starting point for the adoption of heat pumps or electric mobility solutions because adopters are motivated to use their self-generated electricity.

Moreover, there are also some general conclusions for all analyzed innovations. Looking at the relative advantage of these innovations, it stands out, that adopters hardly perceive advantages through the direct functionality of the innovations. For example, heat does not differ qualitatively based on its origin from a heat pump or another heating system. In the same vein, electricity does not differ qualitatively depending on its origin from a PV system or from other sources. Instead, adopters tend to perceive relative advantages through other channels, namely as financial advantages or ecological advantages. Still, both are fraught with insecurities. The financial advantages depend on electricity price developments, as well as on the durability of the implemented innovations. Ecological advantages are often narrowed by impacts of the production of a certain innovation. Both advantages and production impacts are hard to judge or compromised by the feeling that there is a negative ecological impact no matter what decision is made, leading to some sort of perceived helplessness for some adopters. What is left is mostly just a feeling of a positive ecological impact, the expectation of financial advantage or a felt sense of independence, but barely any specific or measurable advantages. Specifically, the utility of smart meters and dynamic electricity tariffs lies within the context of a smart grid. To date, neither change agents nor adopters see sufficient benefits for themselves to overcome the trouble of implementation and possible disadvantages. Because of the crucial role of these innovations in facilitating the control of load within the grid, the support of these innovations in their early-diffusion phases appears to reflect an important challenge.

Another crucial aspect for all innovations seems to be their compatibility, either with the structural context of the building or with each other. The visibility of the different innovations appeared to differ significantly, but for most innovations, suitable digital applications could improve their visibility. For those innovations with a generally higher visibility, their diffusion could give potential adopters the feeling of the technology being well established and therefore more mature. This perception helps with another issue: One major insecurity of the different innovations exists regarding their durability, especially because of high initial costs and the structural context of the building. This attribute cannot be tested, so the trialability cannot support the diffusion. The complexity of different innovations seems to be overcome by a technological background or a technological interest of the adopters.

Moreover, change agents use their potential to reduce the perceived complexity by consulting and through recommendations. Among the different innovations, some relevant groups of change agents appeared to be the same (e.g., installer businesses). Subsidies or regulations, emphasizing the relevance of governance, influence the diffusions of all analyzed innovations. We conjecture that users should be able to combine all analyzed innovations in a useful manner, yet today they tend to adopt them separately. Potential adopters often cannot afford to augment the high costs of a new building by adding the initial costs associated with the simultaneous adoption of multiple innovations. Yet another hindering aspect for the analyzed innovations is a lack of communication and coordination between different change agents of these different innovations. The cooperation of different change agents is generally needed but differs significantly in its quality. Disparate goals, suggestions, and practices in some cases make it harder for these actors to find common ground.

7.5.2 Reflection and Future Research

The chosen research approach was suitable for exploring the factors influencing the diffusion of selected innovations that matter to for the energy transition in Lower Saxony. We conducted five in-depth diffusion studies on these innovations. In doing so, we could also identify groups of actors that play an important role in the diffusion process. The available results on the diffusion of innovations in the context of the energy transition show the importance of different influencing factors on individuals' adoption decisions.

The results provide a qualitative perspective that complements the quantitative modeling and might support the mainly theoretically based models or require amendments in model assumptions. Overall, the derived insights sharpen the simulation models. To this aim, the transition paths of the simulating models were adapted (see Section 8.3). By incorporating the empirical insights into simulation projects, we could thus increase the realism of these models.

Even though many insights could be generated and provide a comprehensive overview of the chosen innovations, there are certain limitations of the results. In line with the whole project, the results relate primarily to Germany. It is not directly possible to infer from this context to others, because Germany-specific conditions, such as feed-in tariffs, influence the results. We also concede some weaknesses inherent to our qualitative and primarily interview-based methodology. Bias could occur due to poorly articulated questions, inaccuracies happen due to poor recall and response bias [77]. Especially, social desirability bias could result from respondents' desire to avoid embarrassment and to project a preferred image to others [120]. However, it is not just the interviewer's perspective that might influence the interviewee. The latter ones' responses might also unknowingly influence the line of inquiry [77]. Another weakness results from the fact that a single interviewer was responsible for the coding of the interviews. It would have been desirable if several authors had coded the same interview and discussed the differences in order to reduce the degree of subjectivity and support the reliability of the results by ensuring inter-coder agreements [78]. Even the selection of an alternative theoretical foundation could have led to different structures within the analysis and therefore could have provided different results.

Apart from these limitations, the results reveal different aspects that should be focused on in future research. Certain innovations such as behavioral adaptions, and certain actors, such as opinion leaders or other groups of adopters, received scarce attention within this study. However, their role within the energy transition is still crucial and should be covered in additional studies. With regard to Rogers' element of time, the presented results concentrate on the diffusion over all adopters, not on the innovation-decision-processes of individual adopters, resulting from the goal to provide information for the transition paths. However, the processes on the micro level appear to be very diverse and deserve further consideration in future research. Interesting for future research could be the choice of other adopters or the extension of the existing sampling to verify the results. A longitudinal approach could investigate and strengthen the results over time. Besides these more general starting points, two specific topics arose from the findings that should receive substantial attention in future research.

First, change agents play an important role in this decision-making process, as they strongly influence the adopter's decision by producing, offering, selling, financing or advising the respective innovations. Over the course of the research project, the question emerged why companies offer certain innovations, even though they do not seem to fit into the existing business model at first. Why, for example, does an energy supplier that earns its money by selling gas offer heat pumps in a contracting model or is involved in the field of electric mobility? The background is increasing pressure, especially due to the energy transition with its trends toward decentralization, decarbonization, and digitization, which has a strong impact on energy suppliers and erodes revenues in the classic energy business. With business model innovations, the energy suppliers try to help shape the transition process toward a sustainable energy supply. These change processes take place differently in each company but are visible everywhere in the energy industry. The best-known examples are the two largest energy suppliers in Germany, E.ON SE, and RWE AG, which have only just split up and are now planning to merge and specialize their business areas [121]. This development seemed unthinkable just a few years ago when both players faced each other solely as competitors. However, the energy transition that continued to gain momentum after the Fukushima reactor accident has reshuffled the cards in the energy industry and swept away old thought patterns. In the future, it will not be sufficient for many companies to make only marginal changes or additions to the existing business model. What is needed are deep business model innovations. The process of change is being accelerated by innovative start-ups that are using digital technologies to build new e-business models, or by an increasing number of new competitors, some of whom do not come from the traditional energy sector.

In the research on these challenges and adjustments that energy suppliers undertake over the course of business model innovations, municipal utilities are hardly considered. These mostly small to medium-sized companies operate in a regionally limited supply area, and local authorities hold the majority of their shares. More than 70 municipal utilities in Lower Saxony (900 in Germany, overall [122]) perform important tasks in the provision of services of general interest. These are the supply of electricity, gas, heat and water as well as the associated grid infrastructure, waste and sewage disposal, and the operation of public transport, street lighting, swimming pools, and parking garages. These companies are particularly affected by the transformation of the energy industry, as they have to develop from locally bound, monopolistic basic suppliers with territorial protection to energy service providers in the free market. They have to deal with increasing competition in their regional home market, decentralization, digitization, rising customer self-sufficiency tendencies, and legal and regulatory requirements, while simultaneously fulfilling tasks of general interest. However, municipal utilities are also shaping the energy transition. The distribution grids, to which 97% of all renewable energy generation plants are connected [122], are often operated by municipal utilities. The sale of PV systems and storage facilities, the establishment of charging infrastructures for electric mobility and energy services such as heat pump contracting can be regarded as new, promising business areas for municipal utilities. The energy transition thus offers opportunities and risks for municipal utilities. How strong the effects of both extremes will be will depend on how well existing business models can be adapted to the changing framework conditions. Identifying the drivers, supporters, and obstacles to business model innovation helps to answer the question of why municipal utilities are adapting or failing to adapt their business model in the context of the energy transition.

Second, the findings show that the analyzed innovations are linked through various relationships to one another, but also to other innovations. Some innovations appear to be in a complementary relationship, others represent a necessary condition for the adoption of another innovation. The influence of the variety of different relationships between innovations on the adoption process or the diffusion of these innovations is yet to be investigated. So far, it can be expected that these relationships might lead to higher levels of complexity and therefore increase the likelihood of non-adoption. However, it might also push the adoption of one innovation along with the adoption of another one. Especially the connection of different innovations or technologies within the context of a smart home is a relevant condition for successful integration of renewable energies. Connecting a number of different innovations within a smart home could lead to a variety of different kinds and degrees of relationships and therefore their impact might be even more substantial. The question, how the relationships of different innovations are perceived by the adopters demands further research. Moreover, it needs to be clarified, in which cases these relationships lead to which outcome for their adoption and diffusion of one innovation individually and a number of linked innovations.

8. From Story to Simulation

J. S. Schwarz, T. Witt

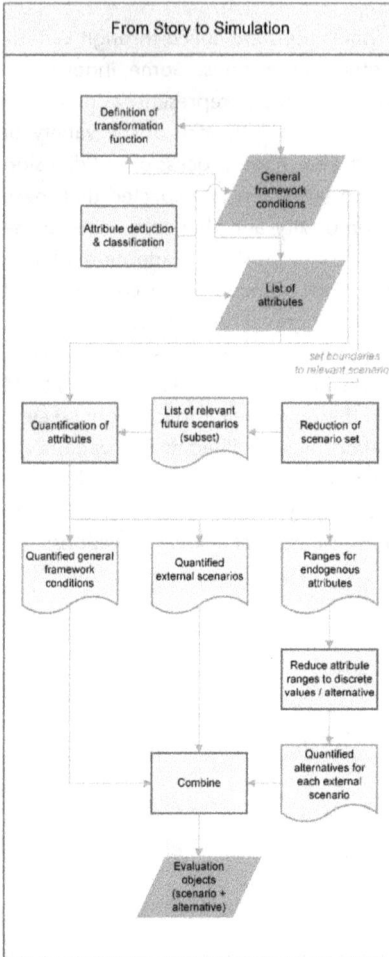

Figure 24: From Story to Simulation in the PDES

The second phase of the PDES aims to close the gap between qualitative future scenarios on the one hand and quantitative modeling and simulation on the other hand (see Figure 24).

First, attributes are defined and quantified as shown in the top of Figure 24 (see Section 8.1). Afterward, alternatives and external uncertainties need to be separated, as a requirement for the application of MCDA. This is shown at the bottom of Figure 24 and explained in more detail in Section 8.2.

In NEDS, not only the year 2050 is considered, but also the transition years up to 2050. These transition paths have also to be quantified, which is described in Section 8.3. To give an example of the developed PDES in NEDS, one future scenario was used for a detailed simulation and evaluation. The selection of this scenario is explained in Section 8.4.

The inputs for this phase are the future scenarios developed via scenario planning as described in Section 6 and the sustainability criteria introduced in Section 0.

The results of this phase are transformation functions, the list of attributes, the general framework conditions, and the evaluation objects.

8.1 Definition and Quantification of Attributes

J. S. Schwarz

The definition and classification of attributes are depicted in the top part of Figure 24. The result of it is a *list of attributes*. This list can be based on two inputs: On the one hand, the qualitative future scenarios can be used to *deduce and classify the attributes*, which are needed to parametrize the simulation and provide input for the sustainability evaluation. On the other hand, the SECs defined earlier can be used as input for the *definition of transformation functions*, which are then used to derive those attributes that are required as inputs for transformation functions. The definition of these transformation functions creates connections between future scenarios, simulation scenarios, and sustainability evaluation. Both sides can be used as input to find missing attributes or SECs. The information model can assist the definition of these transformation functions as described in Section 4.2. By modeling the list of attributes and the SECs in the information model, they can be used for querying and finding suitable connections.

The attributes are classified in one of the categories shown in Figure 7: General framework conditions, scenario-specific framework conditions, endogenous attributes, and derived attributes. These types of attributes are also annotated in the information model. The classification of attributes is depending on the goals and system boundaries of the concrete problem.

Before the quantification of the other types of attributes is done, the scenario set may be reduced based on the general framework conditions, e.g., targets for reducing GHG emissions, which set boundaries for identifying valid decision alternatives. This is necessary because the future scenarios are designed to reflect a broad range of future projections and maybe not all will comply with the general framework conditions. For example, a future scenario with today's shares of renewable energy technologies and increasing energy demand will obviously not meet the GHG reduction targets. However, such a scenario may be used as a reference scenario.

The attributes are quantified based on literature research of related quantitative energy scenarios (e.g., [5]) or other studies. General framework conditions apply to all external scenarios and alternatives and thus are quantified with a single value. Examples could be socio-economic parameters and the lifetime of power plants. The values of scenario-specific attributes represent the external uncertainties and vary from scenario to scenario, but are the same for all alternatives. Examples here could be prices for crude oil and natural gas or the economic growth rate. To capture uncertainty for the decision field within a future scenario, intervals for the endogenous variables are set. Naturally, the process of converting qualitative stories to quantitative assumptions is affected by the subjective assessments of the modeler

[123, 13]. For example, the qualitative story may imply a "good economic outlook". In this case, the modeler's understanding of the term "good outlook" may have an impact on the numerical assumption. Therefore, the assumptions should be documented carefully.

For transparent documentation of the quantification and assumptions, a database for the values and a document with sources and calculations is used. In addition, the dependencies of the quantified attributes should be made transparent to ensure that all models and calculations use the same assumptions. Thus, the information model is used to query these dependencies. The information model is also used to generate the scenario database. A database schema generator has been developed, which automatically creates a schema for a relational database to store the values of attributes and make them available to all models and project partners. A PostgreSQL database is used, which is accessible via SQL and CSV import and export.

In the following, some examples of attributes, their quantification, and the underlying assumptions are shown. Further attributes are described in Section 9 with the descriptions of the simulation models and a list of all attributes can be found in Table 36.

Population (general framework condition):

The population influences the demand for energy, but it is not in the focus in the scenarios and models in NEDS. Thus, it was defined as a general framework condition. The uncertainty of projections for the year 2050 is high and therefore, a medium projection should be chosen.

In the German "Bevölkerungsvorausberechnung" (translated: population prognosis) [124, 125] different projections for the population in Germany until the year 2060 are given. In [5], the "Bevölkerungsvorausberechnung" of the year 2009 [124] is used and the population of Germany downscaled to Lower Saxony based on the principle of solidarity (see also Section 3.2), which means that the proportion of area of Lower Saxony to the area of Germany is used to calculate the number of residents of Lower Saxony:

$$residents\ Germany \cdot \frac{area\ Lower\ Saxony}{area\ Germany} = 70904530 \cdot \frac{4761378\ ha}{35716856\ ha} = 9.452.211$$

The newer "Bevölkerungsvorausberechnung" of the year 2015 [125] assumes a higher population of Germany, which would lead to a population of 9.865.998 in Lower Saxony based on the principle of solidarity. As this would be a difference of 4.38% and the assumptions are highly uncertain, the value from [5] is used.

Economic growth (scenario-specific framework condition):

The gross domestic product (GDP) in percent per year represents economic growth. It measures economic activity and is a significant parameter for energy demand. It mainly depends on the key factor economic structure (KF 7), which provides different projections for the future scenarios. The attribute is defined as external and not influenceable. Thus, it is a scenario-specific framework condition.

The following three studies describe projections for the GDP and were considered for the quantification. In [5], average GDP growth of 0.7% is expected. Because also a decrease in population is assumed, this equates a GDP per capita growth of 1.04%. The World Energy Outlook 2015 [126] contains assumptions, which are subdivided into three periods (2013-2020, 2020-2030, and 2030-2040) with decreasing growth over the years. For the European Union, the assumed average GDP growth is 1.6%. The EU Reference Scenario 2016 [127] contains projections for every country of the European Union. For Germany, the following assumptions for the GDP growth are made [127, p. 133]: Until 2020: 1.3%; 2020-2030: 0.9%; 2030-2040: 0.8%.

Economic growth is one dimension of KF 7. Thus, it has two different projections, which are called slow growth and dynamic growth. As the growth in Lower Saxony and Germany were similar in the past [128], the GDP growth for dynamic growth is assumed in the range of the three shown studies. The GDP growth for the slow growth scenario is chosen slightly lower.

Table 4: Quantification of attribute economic growth (growth of GDP in percent per year).

	Slow growth (Scenarios 2 and 3)	Dynamic growth (Scenarios 1, 4, and 5)
2016-2020	0.6%	1.3%
2021-2030	0.5%	1.2%
2031-2040	0.4%	1.1%
2041-2050	0.3%	1.0%

Distribution of PV systems (endogenous attribute):

The distribution of PV systems is part of the energy mix and is quantified as the share of rooftop PV systems and ground-mounted PV systems. As this attribute is defined as endogenous, for every scenario an interval of possible values has to be defined. The sources for these values are the two energy mixes described in [5]. The first describes a scenario with 100% renewable energies, and the second still contains some fossil-fueled power plants but achieves the 80% GHG emission reduction goal of Germany. For the maximal amount of both types of PV panels [5] contains a potential analysis, which was taken into account for the intervals.

The attribute depends on multiple KFs from the scenario planning (see Section 6.1):

- **Topology of distributed energy resources (DER) (KF 2):** This KF determines the proportion of local DER (focus on rooftop PV systems) and central large power plants (focus on ground-mounted PV systems).
- **Energy mix (KF 5):** This KF determines the share of different energy sources. Thus, it determines the proportion of PV systems in the total energy mix.
- **Demand for energy (in private households) (KF 6) and economic structure (KF 7):** These KFs determine the demand for energy, which has to be met by the energy mix.

For the intervals, the following assumptions are taken. Rooftop PV systems feed into the grid on low voltage level and ground-mounted PV systems on medium voltage level. As the energy mix bases upon the scenarios in [5], generation can only be shifted between the different types of PV systems and not to other energy resources and the potential analysis of [5] has to be taken into account for these shifts.

Table 5: Quantification of attribute distribution of PV systems (in percent of total energy mix).

Scenario	Share of rooftop PV systems	Share of ground-mounted PV systems	Comment
1	4.43%	1.48%	KF 5 defines the energy mix as the status quo. Thus, the percentage of 2015 is used.
2	25.39 – 28.56%	24.81 – 27.98%	KF 5 describes a high amount of renewable energies. Thus, the 100% renewables scenario [5] is the base. KF 2 defines a focus on small local energy resources. Thus, the share of rooftop PV systems is comparable to ground-mounted PV systems.
3	3.36 – 22.61%	20.23 – 39.48%	KF 5 describes some fossil backup plants. Thus, the 80% GHG emission reduction scenario [5] is the base. KF 2 defines a focus on distributed large renewable power plants. Thus, the share of ground-mounted PV systems is higher.
4	9.52 – 15.87%	37.50 – 43.85%	KF 5 describes a high amount of renewable energies. Thus, the 100% renewables scenario [5] is the base. KF 2 describes a focus on small distributes energy resources. Thus, the share of ground-mounted PV systems is higher.
5	25.39 – 28.56%	24.81 – 27.98%	KF 5 describes a high amount of renewable energies. Thus, the 100% renewables scenario [5] is the base. KF 2 defines a focus on small local energy resources. Thus, the share of rooftop PV systems is comparable to ground-mounted PV systems.

The values for the attributes are imported to the scenario database via a CSV file as shown in Figure 25. In this file, the columns (A-C) specify the attribute; column (D) specifies the type of attribute. Then the value (E) and its unit (F) are specified and in the last three columns (G-I) the scenario, alternative, and year of the value are specified. For general framework conditions, one value for all scenarios and alternatives is imported. For scenario-specific framework conditions, a value for every scenario is defined. For endogenous attributes, the intervals of possible values for all scenarios are defined. The values in the scenario database can be filtered by using SQL to get only the needed values, e.g., for a specific scenario and alternative.

	A	B	C	D	E	F	G	H	I
1	domain	domainobject	name	type	value	unit	scenario	alternative	year
2	User	Sociodemography	Population	GeneralFrameworkCondition	9,452,211	all	all		2050
3	Market	Economy	EconomicGrowth	ScenarioSpecificFrameworkCondition	1.3 %		Scenario1	all	2020
4	Market	Economy	EconomicGrowth	ScenarioSpecificFrameworkCondition	1.2 %		Scenario1	all	2030
5	Market	Economy	EconomicGrowth	ScenarioSpecificFrameworkCondition	1.1 %		Scenario1	all	2040
6	Market	Economy	EconomicGrowth	ScenarioSpecificFrameworkCondition	1 %		Scenario1	all	2050
7	Market	Economy	EconomicGrowth	ScenarioSpecificFrameworkCondition	0.6 %		Scenario2	all	2020
8	Market	Economy	EconomicGrowth	ScenarioSpecificFrameworkCondition	0.5 %		Scenario2	all	2030
9	Market	Economy	EconomicGrowth	ScenarioSpecificFrameworkCondition	0.4 %		Scenario2	all	2040
10	Market	Economy	EconomicGrowth	ScenarioSpecificFrameworkCondition	0.3 %		Scenario2	all	2050
11	Market	Economy	EconomicGrowth	ScenarioSpecificFrameworkCondition	0.6 %		Scenario3	all	2020
12	Market	Economy	EconomicGrowth	ScenarioSpecificFrameworkCondition	0.5 %		Scenario3	all	2030
13	Market	Economy	EconomicGrowth	ScenarioSpecificFrameworkCondition	0.4 %		Scenario3	all	2040
14	Market	Economy	EconomicGrowth	ScenarioSpecificFrameworkCondition	0.3 %		Scenario3	all	2050
15	Energy	EnergyMix	RooftopPV	EndogenousDomainObjectAttribute	4.43 %		Scenario1	all	2050
16	Energy	EnergyMix	FreeFieldPV	EndogenousDomainObjectAttribute	1.48 %		Scenario1	all	2050
17	Energy	EnergyMix	RooftopPV	EndogenousDomainObjectAttribute	25.39 - 28.56 %		Scenario2	all	2050
18	Energy	EnergyMix	FreeFieldPV	EndogenousDomainObjectAttribute	24.81 - 27.98 %		Scenario2	all	2050
19	Energy	EnergyMix	RooftopPV	EndogenousDomainObjectAttribute	3.36 - 22.61 %		Scenario3	all	2050
20	Energy	EnergyMix	FreeFieldPV	EndogenousDomainObjectAttribute	20.23 - 39.48 %		Scenario3	all	2050
21	Energy	EnergyMix	RooftopPV	EndogenousDomainObjectAttribute	9.52 - 15.87 %		Scenario4	all	2050
22	Energy	EnergyMix	FreeFieldPV	EndogenousDomainObjectAttribute	37.50 - 43.85 %		Scenario4	all	2050
23	Energy	EnergyMix	RooftopPV	EndogenousDomainObjectAttribute	25.39 - 28.56 %		Scenario5	all	2050
24	Energy	EnergyMix	FreeFieldPV	EndogenousDomainObjectAttribute	24.81 - 27.98 %		Scenario5	all	2050

Figure 25: Quantified attributes in CSV for import in scenario database

8.2 Development of Alternatives

T. Witt

After having separated endogenous and exogenous attributes, different alternatives capturing the uncertainty for the decision field within a future scenario (see Section 6.1.1) need to be specified. In the case of power generation systems, alternatives could be different system configurations. For example, a system configuration can represent the overall installed capacity of different renewable energy technologies (such as wind energy, solar energy, or bio-energy). Modeling the performance of alternatives in different scenarios via energy system analysis can then be used as input for evaluation via multi-criteria analysis.

The previously defined intervals for endogenous attributes are used in the *quantification of endogenous attributes* to define concrete values for the endogenous variables that represent, according to our definition, the alternatives.

The consistency analysis from scenario planning (see Section 6.1.4) can help to define concrete values in a structured way. Although this method has also been used to define future scenarios, the difference is that, here, the *endogenous attributes* are used *as KFs*, so that (up to four) projections are defined for the endogenous attributes. Based on the consistency of the projections, *k*-means cluster analysis can be performed to generate *k* alternatives, which are described with qualitative stories. To convert these storylines into quantitative assumptions, the subjective assessments of the system modeler again play a role, since there is no objective way to select specific values for endogenous attributes [123, 13].

In this project, we selected the following KFs to describe the alternatives:

(1) Distribution of PV systems
(2) Distribution of wind energy power plants
(3) Share of households participating in energy management
(4) Diffusion of storage systems connected to the high and medium voltage levels of the power grid
(5) Diffusion of storage systems connected to the low voltage level of the power grid
(6) Diffusion of information and communication technology (ICT) standards in the power grid.

Applying consistency and cluster analyses lead to three alternatives (see Table 6):

- **Alternative 1 (A1):** Local power generation with flexible energy demand, which has a focus on onshore wind power, rooftop PV systems, and small-scale storage systems (*"decentral"*)
- **Alternative 2 (A2):** Large-scale storage systems and power generation, which has a focus on offshore wind power, ground-mounted PV systems, and large-scale storage systems (*"central"*)
- **Alternative 3 (A3):** Middle ground: a mix of all generation and storage technologies (*"middle ground"*)

Table 6: Qualitative description of alternatives for the power generation system of Lower Saxony

	Local power generation with flexible energy demand	Large-scale storage systems and power generation	Middle ground
Distribution of PV systems	Focus on rooftop PV systems	Focus on ground-mounted PV systems	No focus
Distribution of wind energy power plants	Focus on onshore wind energy	Focus on offshore wind energy	No focus
Share of households participating in energy management	High share	Low share	Average share
Diffusion of storage systems connected to the high and medium voltage levels of the power grid	Many small storage systems	Few large-scale storage systems	Mixed diffusion
Diffusion of storage systems connected to the low voltage level of the power grid	More storage systems	Less storage systems	Average diffusion
Diffusion of ICT-standards in the distribution grid	High diffusion	Low diffusion	Average diffusion

Having defined the three alternatives in a qualitative way, the next step is to quantify them with concrete, numerical assumptions. The intervals from the first quantification step are the base for this quantification of the alternatives. As already shown in the previous Section for the share of PV systems, also the share of wind engines is based on the scenario in [5]. (If certain attributes are not specified in [5], we used own assumptions.) Some examples of the quantified assumptions are shown in Table 7. Further assumptions for each simulation and optimization models are detailed in the individual models' Sections (see Section 9). The values for the attributes are stored in the scenario database and replace the intervals, which were imported in the first quantification step.

Table 7: Quantified attributes of alternatives for the power generation system of Lower Saxony

Attribute	Central	Medium	Decentral
Share of rooftop PV systems	3.36%	12.98%	22.61%
Share of ground-mounted PV systems	39.48%	29.86%	20.23%
Share of onshore wind energy power plants	26.36%	28.34%	30.31%
Share of offshore wind energy power plants	20,17%	18,19%	16.22%
Storage systems connected to the high and medium voltage levels	12.33TWh	8.22TWh	4.11TWh
Storage systems connected to the low voltage level	4.4GWh	4.9GWh	5.39GWh

8.3   Transition Paths

T. Witt, J. Minnemann, M. Kleinau

In this project, the future scenarios and alternatives up to now have been defined only for the year 2050 *(target state).* To investigate, how the transition of Lower Saxony's power generation system until 2050 can be made *(How can the system evolve so that it reaches the target state?),* we define one *transition path* per alternative and scenario. Note that we assume strong path dependency (using the concept of [129]), so that, if a path is entered in the year 2020, the path cannot be left and will be followed through until 2050. The transition paths describe the transition of Lower Saxony's power generation system in a simplified way, inasmuch as only the years 2015, 2050, 2030, 2040, and 2050 are considered for quantitative modeling. This simplification is made to keep both the effort for quantification of scenarios and alternatives as well as the overall runtime of the simulation and optimization models in acceptable limits. Additionally, as most model assumptions and results are subject to uncertainty, performing quantitative modeling and simulation for each year until 2050 would suggest that the results are accurate and reliable. However, since scenarios usually only cover a small fraction of the future state space, only general tendencies and what-if statements can be made.

In Figure 26, a simplified illustration of the transition paths is shown as a so-called *transition graph,* which is a directed graph where each vertex (circle) represents one alternative in a specific year, and the edges represent possible transitions.[1] The different colors indicate different shares of energy conversion technologies for each alternative. Accordingly, to quantify the transition paths, all endogenous attributes, scenario-specific parameters, and general parameters need to be defined for all considered years. This increases the effort for quantification of assumptions, compared to modeling only the target state in 2050. For example, in this project, 3 alternatives have been defined. Therefore, in total, $3\ (alternatives) \cdot 4\ (years) \cdot 5\ (scenarios) + 1(status\ quo) = 61$ consistent parameter combinations have to be defined for quantitative modeling and simulation. To keep the effort for quantification in an acceptable limit, in our first approach, we used linear extrapolation between 2015 and 2050 to quantify the assumptions for each transition path in each year.

---

[1] Note that other future scenarios are not depicted in Figure 26. For each future scenario, one transition graph needs to be specified, because each scenario represents a different section of the future state space.

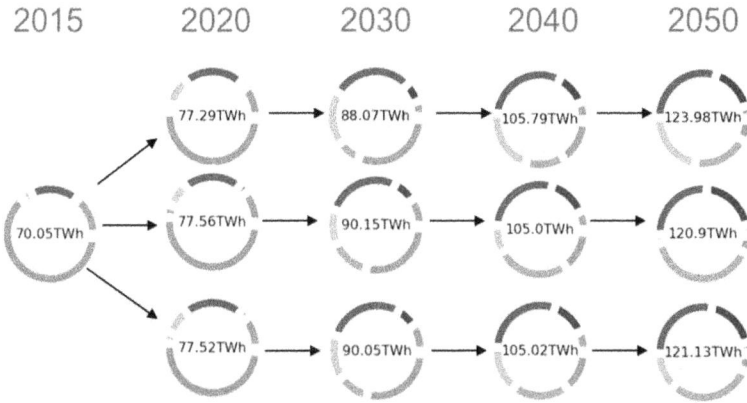

2015   2020   2030   2040   2050

Figure 26: Simplified illustration of transition paths. Pie charts depict the annual energy production and the energy mix (grey: fossil fuel, green: biogas, dark blue: wind offshore light blue: wind onshore, yellow: PV-rooftop, orange: ground-mounted PV)

This concept of transition paths implies that, in effect, alternatives and scenarios are defined for all of the above-mentioned years, so that a multi-criteria evaluation of these paths' sustainability also needs to be done for each year, before the evaluation can be aggregated for a transition path. Transition paths consist of alternatives, which makes the concepts prone to misunderstanding. To avoid confusion and allow for differentiation between alternatives and paths, the paths are named as follows:

- Path 1: Path toward Alternative 1 ("decentral")

- Path 2: Path toward Alternative 2 ("central")

- Path 3: Path toward Alternative 3 ("middle ground")

Having Rogers' concept of diffusion of innovation [65] with hindering and facilitating context factors in mind, the diffusion of technologies usually does not proceed in a linear, but rather in a sigmoidal way. Therefore, the assumptions for selected attributes were adjusted slightly, based on the empirical investigation described in Section 7.

For these adjustments of the transition paths, the qualitative results of the diffusion studies and the assumptions of the simulation models were brought together and adapted. In detail, the assumptions of the simulation models regarding PV systems with storage, smart meters, electric mobility, and heat pumps were empirically embedded based on the results of the diffusion studies (see Section 7.4). As a result, the assumed linear course in the parametrization of simulation models was adjusted.

The sequence of such an empirical embedding of the simulation attributes will be shown in detail by the example of the smart meter. The results of further adjustments are then briefly summarized for Scenario 3.

The attribute *measurement systems in households* distinguish between an intelligent measurement system with the gateway (smart meter) and a digital measurement system that does not have a gateway and therefore does not send any data (digital meter). Since the rollout of smart meters is legally formulated, this provides the framework for its diffusion and was thus defined as an attribute. However, since the minimum requirements for the smart meter rollout have not yet been met (see Section 7.4.2), the assumptions of the diffusion have been slowed down and shifted more into the future. The cost of a smart meter is higher than the cost of a mechanical electricity meter. Distribution system operators, therefore, do not see the need for a rollout to customers who are not covered by legal requirements. Hence, the distribution of smart meters and digital meters have been changed in such a way that the digital meters are implemented where the use of the smart meter is not intended. Digital meters are already being used and have therefore become more widespread than smart meters. The diffusion of digital meters will also take a long time and will only gradually replace mechanical electricity meters. The reason for this is that built-in mechanical electricity meters are usually only replaced by distribution system operators at the end of their service life. Beyond the year 2030, mechanical electricity meters will, therefore, continue to exist in households and small companies. A slower diffusion of smart meters, a comparatively stronger diffusion of digital meters and a residual stock of mechanical electricity meters led to an adjustment of the attribute and thus to a sharpening of the simulation assumptions through empirical embedding.

In addition to the described adjustments to the *smart meter*, further adjustments have been made. The compatibility of the structural context plays a decisive role in the diffusion of the *heat pump*. Because it is mainly used in new buildings, but not in existing buildings, the diffusion process is taking place slowly. This slow course leads to a reduction of the diffusion and thus to an adjustment of the assumptions in the attribute. The assumptions on the diffusion of *electric mobility* were also reduced based on the empirical results of the diffusion study. The main reason for this is the expected failure to reach the target of one million electric cars on German roads by 2020, which has been announced by the Federal Government. The slow diffusion of electric cars is reinforced by a charging infrastructure that does not yet cover the whole country. The combination of *photovoltaics and storage* is characterized by interwoven diffusion processes. The knowledge gained about the diffusion of photovoltaics does not allow a clear prognosis. For the diffusion of storage systems, especially the influences of the retrofitting of existing PV systems, which are going to fall out of the feed-in tariff, were taken into account.

8.4   Selecting Future Scenarios for Simulation

J. S. Schwarz

After the quantification of all attributes, the next step is the simulation. Five scenarios were developed to represent external uncertainties with the goal to find strategies in the endogenous part of the system that would be beneficial in all external circumstances. Finally, only one scenario was simulated with all simulation models, due to lack of time. Therefore, the decision which scenario should be simulated first turned out to be very important. Scenario 3 was chosen to be the first simulated scenario mainly due to its assumed energy mix. In this scenario, the energy mix relies mainly on renewable energies and the 80% GHG reduction goal is fulfilled as described in [5], but some gas power plants are used as a flexible backup. This scenario best matches the current progress of climate protection described in the following studies. The "Klimaschutzbericht 2018" [130] states that the climate protection goals of the German government for the year 2020 will most likely not be fulfilled. Also, [131] describes a benchmark scenario, which extrapolates the GHG emission reduction between 1990 and 2015 up to the year 2050. In this benchmark scenario, a GHG reduction of 61% would be achieved. Thus, achieving the 80% GHG reduction goal will need more activities of the government and the 95% GHG reduction goal even more.

9. Modeling and Simulation

M. Nebel-Wenner, C. Reinhold, J. S. Schwarz, F. Wille

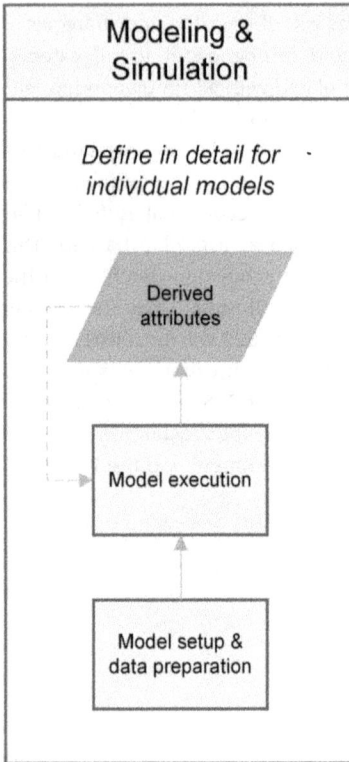

Figure 27: Modeling and simulation in PDES

Modeling and simulation are used to enrich the quantified future scenario (see Section 8) with additional quantitative data for the derived attributes (see Section 4.2), which are used in the evaluation (see Section 11). This is necessary in order to integrate aspects in the evaluation, which are not covered by existing studies. It also allows considering new technologies and approaches via integrated modeling of multiple domains and system levels. In the PDES modeling and simulation is only described rudimentarily (see Figure 27). The steps model setup & data preparation and model execution have to be done for all simulation models, but may have individual requirements. Thus, they have to be defined in detail for the individual models.

The transformation of an energy system toward sustainability encompasses adaptions on different levels of the system. The objective of the modeling is an adequate representation of the system to be considered with regard to different sectors and a modular and expandable structure of the holistic simulation environment. An energy supply system can be described on two levels:

On a micro-level, we include an analysis of possible behavioral adaptions (Section 9.1) and technical adaptions in buildings (Section 9.2) as well as grid control in distribution grids (Section 9.3). Section 9.4 focuses on the coupling of the two sub-models Building and Smart Grid and the following Section 9.5 the associated simulation scenarios and their execution methodology. On a macro-level, technical adaption regarding grid planning on medium and higher voltage levels are considered (Section 9.6 and Section 9.7) as well as economic aspects of the general economy and the energy sector (Section 9.6 and Section 9.8) are analyzed. Additionally, a Life-Cycle Assessment (LCA) approach is presented in

Section 9.9 in order to assess the environmental and in parts the social sustainability of the electricity generation system. Section 0 describes the coupling of the models on the macro-level.

Table 8: Overview of used Simulation Models with temporal resolution and level

Simulation Models	Level	Temporal resolution
Building with user behavior - *eSE*	Micro	
Smart Grid - *ISAAC*	Micro	Minutes / Hours
Electricity Market - *INES*	Macro	
General Economy - *CGE*	Macro	Years
Power Grid Planning	Macro	Years of transition

Quantitative and qualitative models are used, which differ considerably in terms of the methods used, the subject of investigation, degree of detail and temporal resolution (see Table 8 for an overview). In order to investigate and address the questions in a holistic manner, the models are connected and integrated into a simulation framework, which is shown in Figure 28.

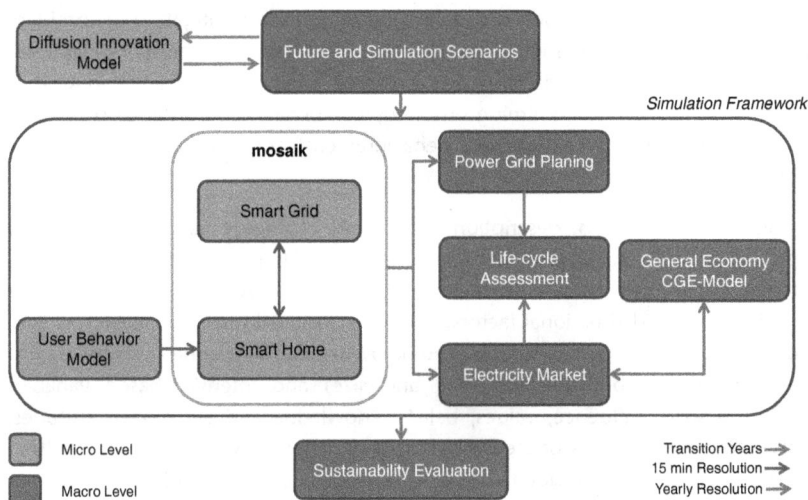

Figure 28: Models and Simulation Framework

The co-simulation framework mosaik allows to couple of simulation models, which can represent different domains and can be implemented in different programming languages and paradigms [26, 132]. This coupling allows using existing models in a new context and avoids time-consuming implementation of new models. Because of the different temporal resolutions of the models in NEDS and the different modes of data exchange between them, not all simulation models are integrated into a mosaik simulation. The building model and the smart grid model need close coupling, because they heavily depend on each other's data. Therefore, these models were coupled in mosaik (see Section 9.4 for a detailed description). Theoretically, all models could be coupled in mosaik, but to optimize the execution time of the whole simulation framework, the other simulation models are coupled by CSV files.

In the following Sections, the models' structures and methodologies will be explained, as well as the coupling between them. The boundary conditions and datasets for the simulations will also be presented.

9.1   Theoretical and Empirical Analysis of User-Behavior

F. Wille

One key factor in the transformation of the Lower Saxony energy scenario is the energy mix. With an increased amount of renewable energy resources, fluctuations in energy generation increase and consequently a problem of possible divergences between energy supply and demand arises. This assumed change in the energy system can ensue adaptions on different levels, one of them being the behavioral level of energy users. To model and assess the potential for mitigating the problem of discrepancies between supply and demand by shifting user behavior, the user-behavior model needs to describe the timing of energy use behavior as well as factors, which influence the timing of energy use behavior, in order to be able to deduce points of intervention and behavioral costs of shifting an energy use behavior in time.

To deduce interventions, descriptions or models of energy use behavior should depict factors influencing user behaviors. Frederiks, Stenner, and Hobmann [133] differentiate in their review on factors influencing household energy behavior individual-level and situational factors. Focusing on individual-level predictors they find that sociodemographic variables (home ownership, household- income, family size and composition, dwelling type and size) and intention-based variables (normative social influence, values, beliefs, knowledge and awareness, attitudes, goals, and motives) are associated with household energy use behavior. The pattern of association, however, is not always clear, especially in the case of intention-based variables [133]. This problem, as well as a neglect of contextual factors in explaining energy use behavior, is echoed by [134].

Thus, this analysis first aims at describing energy use behavior and identifying contextual factors, which influence the timing of energy use behavior. Theoretical assumptions of the user-behavior model are based in behavioral theory, which states that contextual factors influence behavior, because the context sets the consequences, which select behavior, and delivers discriminative stimuli, which indicate different structures of contingencies [135]. Based on this assumption interventions are theoretically deduced. In a second analysis step, behavioral adaptive costs for shifting energy use behavior under the identified contextual restrictions are empirically estimated to give information on flexibility potential in energy demand attributable to user behavior.

9.1.1 Description of User Behavior

To arrive at a detailed description of behavior including energy use behavior and to identify contextual factors, we analyzed data from the latest German Time Use Survey from 2012/2013 [136]. The survey data comes from a representative quota sampling procedure from German private households with participants from the age of ten being eligible. For three days, including weekdays and weekend days participants' activities were collected in ten-minute intervals via an activity diary. Information about 5040 private households with 11371 individuals was recorded. The diary data were analyzed and categorized into about 165 activities [137].

Table 9: Summary of Time Use Data into 22 activities [138]

Description of activity	Code number[1]
sleeping	11
physiological recreation like food and drink consumption and washing oneself	12, 13
occupational activities	2
education and further education in school, college or at the university	31-34
other education-related activities like homework, studying	35, 36
preparing meals and cleaning up afterwards	41
chores at home	42
doing laundry, mending textiles	43
gardening and animal care	44
handicraft activities	45
shopping and use of services not at home	46
childcare at home	47
care and support of adult household members	48
other housekeeping and support activities for the family	49
volunteer work	5
social activities and cultural entertainment	6
hobbies, sports, game playing	7
reading	81
watching TV, DVD, etc.	82
listening to radio and music	83
using a computer or smartphone	84
travel and commute activities	9

Note [1] Original code number from Time Use Survey [139]

A model which identifies and describes different behavior patterns can be built by grouping the activities for weekdays and weekend days (Saturdays, Sundays and national holidays) according to their similarity in the pattern of activities over a day. From the more than 165 hierarchical organized activities, we used 22 upper categories (Table 9) to cluster the different behavior patterns. Since participants filled out two weekdays or two weekend days, we randomly selected one data entry for weekdays and one entry for weekend days. So that for weekdays we had a total of $n = 10589$ subjects to be clustered and $n = 10654$ for weekend days. The distance between the subjects is measured using the Levenshtein Distance [140] and clustered based on the Partitioning around Medoids method (PAM) [141] in R (PAM package cluster version 2.0.6).

Based on the validation criterion average silhouette width with a possible range from $-1 \leq s_i \geq 1$ [141] (Figure 29), we chose a three groups cluster-solution for the weekday data and a six group cluster solution for the weekend data (the five-cluster solution had two clusters with comparatively very high but also low separation values and the seven-cluster solution had two relatively small sample sizes). The average silhouette width values are positive, but very close to zero, indicating that within-cluster cohesion is only slightly larger than cluster separation.

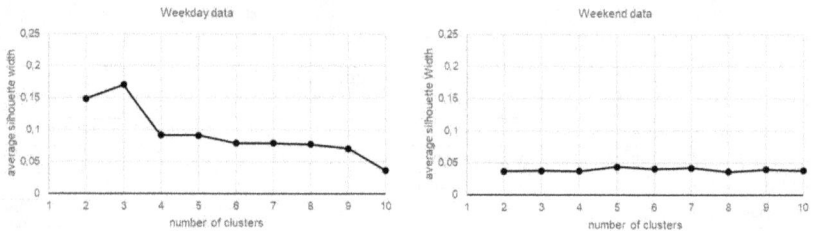

Figure 29: Selection criterion average silhouette width [138]

For an overview of cluster sizes, view Table 10. As can be seen, the clusters are large enough in order to be useful for simulation purposes.

Table 10: Cluster size for weekday and weekend

Day Type	Cluster 1	Cluster 2	Cluster 3	Cluster 4	Cluster 5	Cluster 6
Weekday	4235	1991	4363	-	-	-
Weekend	1278	1946	2482	1260	2738	950

The identified behavior patterns for the weekday data can be viewed in Figure 30 and Figure 31, the behavior patterns for the weekend data can be viewed in Figure 32 through Figure 34.

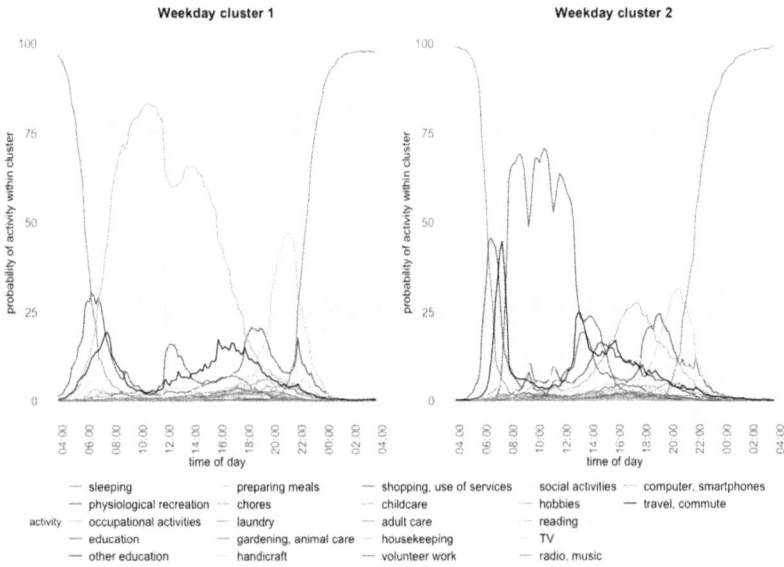

Figure 30: Behavioral activity patterns for weekday data in cluster 1 (n = 4235) and in cluster 2 (n = 1991) [138]

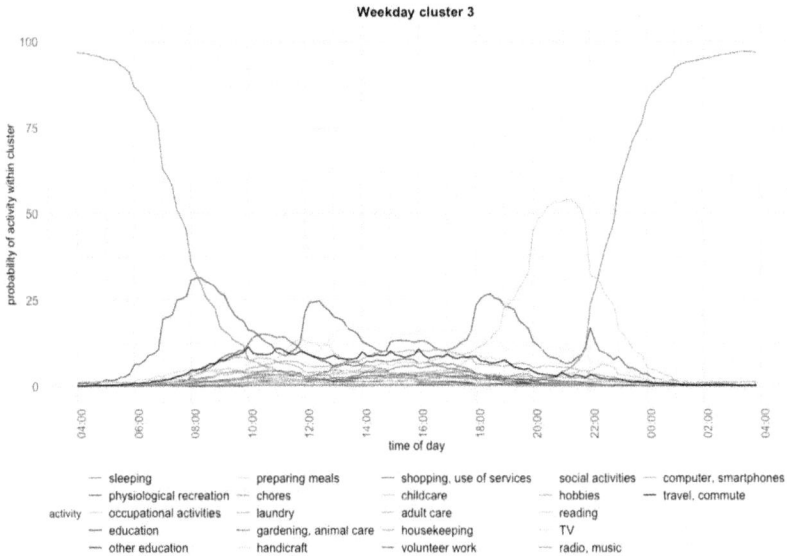

Figure 31: Behavioral activity patterns for weekday data in cluster 3 (n = 4363) [138]

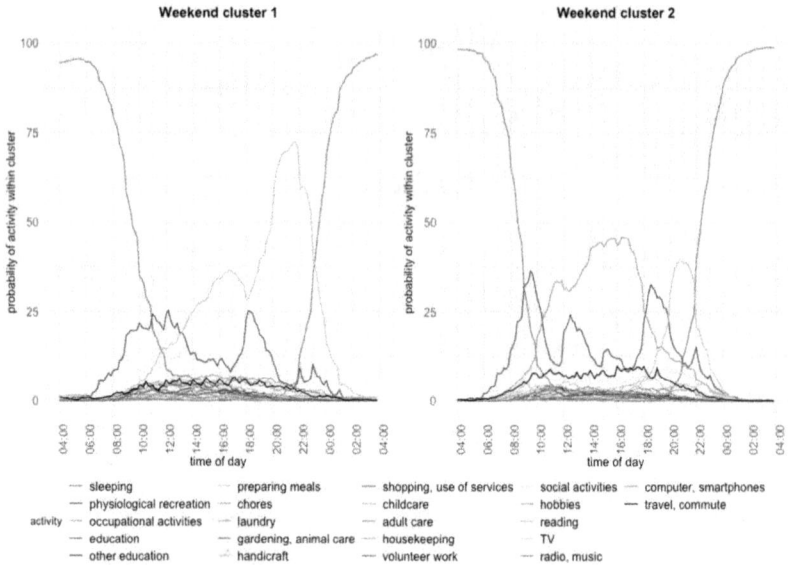

Figure 32: Behavioral activity patterns for weekend data in cluster 1 (n = 1278) and in cluster 2 (n = 1946) [138]

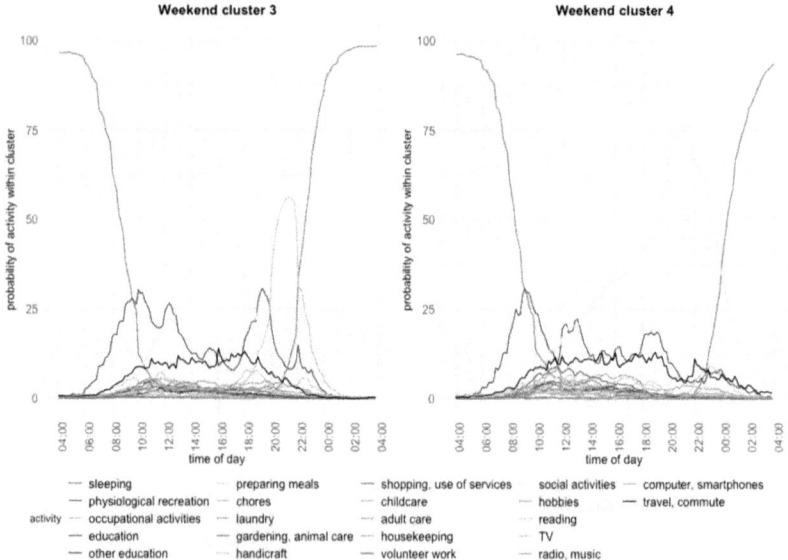

Figure 33: Behavioral activity patterns for weekend data in cluster 3 (n = 2482) and in cluster 4 (n = 1260) [138]

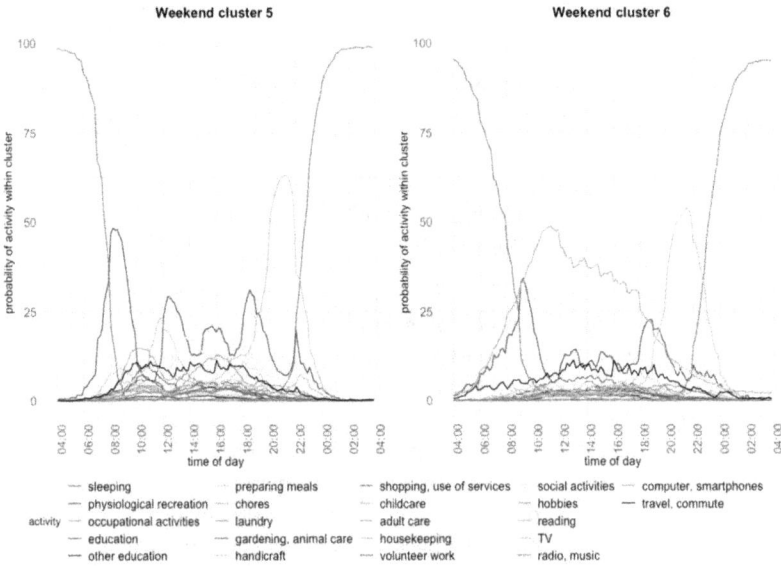

Figure 34: Behavioral activity patterns for weekend data in cluster 5 (n = 2738) and in cluster 6 (n = 950) [138]

As can be seen in the figures above, the main difference between the clusters arises from differences in the timely distribution of certain activities. For weekday data, those are occupational and educational activities and their absence in weekday cluster 3, as well as differences in slope and beginning rises and declines in the probability of the sleeping activity. In shape, the activity sleeping curve is very homogeneous for all weekday and weekend clusters. The main differences between the weekend data stem from the sleeping activity, occupational activities (weekend cluster 6), hobbies (weekend cluster 2), social activities (weekend cluster 3 and 4) and watching TV (weekend cluster 1 and 5). Looking at the social activities and watching TV, it becomes clear, that what mainly differentiates the clusters is not only the overall probability of this activity within a cluster but their timely distribution over the course of a day. This result, of just a few different clusters, three for weekdays and six for weekend days, with main differentiating activities for many different behavioral activities in 144 ten-minute time slots, points toward a relative homogeneous structure of those behavioral activities. If this description has merit, one should further analyze how those differentiating activities structure other behavioral activities, such as energy-related behaviors in households.

9.1.2 Interpreting Differences between Activity Patterns as Caused by Contextual Factors

Morris [142] identifies two meanings of *context*. The meaning of context-as-history and the meaning of context-as-place, as it is often employed in social, cognitive and behavioral sciences. According to Morris [143] context-as-place may be most usefully employed if restricted to either a formal meaning, as an initial or boundary condition or a functional meaning, as conditions that alter functional relations within the three-term-contingency. For this analysis, both meanings are useful.

In a formal meaning, the identified differentiating activities between clusters restrict and structure the possible times of energy appliance use. Occupational and educational activities, and in most cases also hobbies and social activities are associated with absence from home. This absence renders using an electrical appliance impossible or more unlikely since its use during absence would require the availability and employment of a programmable timing function or internet-based application. Thus, the first argument for interpreting the above stated differentiating activities as contextual factors is that showing one of those behaviors excludes the possibility of showing a behavior at home and thus limits the possible hours within the day where a home-associated appliance use behavior can be shown. This limitation is viewed a as restriction. This appears sufficient to fulfill the formal meaning of a contextual factor.

In cases, where hobbies or social activities are performed at home or for the other differentiating activities like sleeping and watching TV, one must argue in how far those activities can function as context factors by selecting behavior and through this arrangement restricts the variability of energy-related household behaviors. For our argument (and building model), we assume linkages between certain activities and use of an electrical appliance (view Section 9.2.3) and consider those activities our energy-related household behaviors. If a hobby or social activity is performed at home one can say, that it is performed in preference to other behaviors, as for example energy related household behaviors. Choosing one activity over another would alter the timely distribution of other behaviors and fulfill the criterion of restricting the variability of behavior. Since this would hold for all cases of choice behavior, it is not useful in defining a meaningful category of a contextual factor. However, preference is also associated with a specific time point, which is determined by differential consequence outcomes when operating on the context at that point of time versus another point in time. It thus seems useful to interpret behavioral activities as restricting contextual factors if they influence the variability of other behaviors and if they correspond to regular occurring changes in the available consequences. In our case, the differentiating activities between clusters seem to correspond to regularities determined by day-and-night rhythm (sleeping, occupational and educational activity, hobbies and social activities) and by societal structures (working and schooling hours, TV program, sleeping, hobbies and social activities). If we observe regularities in the environment, which are associated with similar changes in consequences when operated upon by a large number of individuals or relevant subgroups, we can categorize those as contextual factors (for a detailed version of this argument refer to [138]).

If this interpretation is justified, one has also identified points of intervention for changing behavior. These interventions can target changing the regularities, which set the consequences for behavior. This seems a good starting point, especially in cases where the problem is not one of net energy supply but one of matching supply and demand at certain time points. In practical terms, one could think of increasing the flexibility of working and schooling hours during the weekdays. On the one hand, those types of interventions, which aim at changing the structure of producible consequences as determined by the regularities of a context factor, are more difficult to implement than interventions targeting for example only a specific consequence, like incentives. On the other hand, they broaden the perspective beyond usually discussed interventions such as laws and regulations, commercials, available technologies and incentives [144]. Evaluating the effect of such possible interventions would be an interesting step, once the user model has been built, but is not part of the current project.

In sum, what becomes evident through this analysis is, that if we want to specify a behavior-theoretical user model (which in future work might lend itself to evaluating effects of context interventions), we need to consider contextual factors for energy-related user behavior. In order to evaluate the possibilities for shifting user behavior and its potential in mitigating the problem of discrepancies between energy supply and demand, empirical estimations of *user flexibility* should consider these contextual restrictions.

9.1.3 Estimation of Behavioral Adaptive Costs

To build the user-behavior model, we described the timely distribution of energy-related household behaviors in ten-minute intervals for three weekday clusters and six weekend day clusters representing the main structuring context factors. As one aim is to consider *user flexibility* within the user model, we conducted a correlational study to empirically describe an indicator for user flexibility, which we call behavioral adaptive cost (*bac*) as it empirically estimates the effort for shifting the use of an electrical appliance in time.

Questions regarding the user's role in offering flexibility within a renewable energy system target two main aspects, the role in the demand for energy and in the supply of energy [145]. For this analysis, we focus on the demand side and the question of load shifting, even though in principle when it comes to realizing a sustainable energy system, measures toward energy-demand reduction and furthering energy efficiency are also important. Though different definitions of demand response exist, a common theme is, that it reflects electricity demand, which is responsive (flexible) to economic signals [146]. According to Schuitema, Ryan and Aravena [145], studies on shifting loads by introducing time-of-use tariffs realize an energy shift from consumption peak times to off-peak times by approximately 8%. A qualitative study [147] on people's daily interactions with energy-consuming products and systems (including shifting behavior) emphasizes a general inflexibility in respondents' willingness to change their interactions with a wide variety of everyday energy consumption products. In light of the assumed importance of structuring context factors for household energy use behavior, such a limit in pricing interventions seems coherent. Without lifting context restrictions, we assume the potential of shifting energy use behavior in time to be limited. Nonetheless, analyzing in how far user flexibility can be incorporated in a building model and smart-grid model to address and evaluate such questions is relevant, because it helps to differentiate possibilities for load shifting in the residential sector.

While it is relatively clear on a technical level, what user flexibility within a smart grid system should achieve, namely altering electrical load patterns by means of changing behavior of humans, it is important to look at conceptualizations in behavioral terms to understand the theoretical idea behind the bac indicator for user flexibility. In general, *behavioral flexibility* describes an organisms' adjustment of behavior to changing environments throughout its life [148]. In experimental psychology and behavioral ecology, it is studied within different study designs, like for example reversal learning, set-shifting, self-control and problem-solving, leading to different conceptualizations of *behavioral flexibility* [149]. Bond, Kamil and Balda [150] also identify at least three different, though similar connotations of the term flexibility within the behavioral literature: In a first sense, flexible organisms modify their behavior quickly based on limited experience in response to subtle variations in consequences or context. Secondly, the term flexible is used to refer to exploratory, playful and versatile behavior without changing contexts and third it refers to behavior patterns, which can be repeatedly reversed depending on changes in context, as it is studied within the operant procedure of reversal learning by reversing reward contingencies. The way, user flexibility is viewed within demand response definitions, is limited in perspective compared to a general concept of behavioral flexibility as adjusting to changing environments. The changing environmental aspect within a renewable energy system is the fluctuating energy resource, which is not passed on directly to the (residential) user for example in form of non-availability of energy at certain time points. Within our future energy scenarios, the smart-grid and other information communication technology regulate supply and demand in a way that a user adjusts to a fluctuating, but not a depletable resource. Not part of our investigation, but in line with the demand response idea of user flexibility, this is achieved via pricing arrangements that may differ in how directly they signal the state of the resource. This approach leaves the main consequence or function of operants requiring electric energy unchanged. Whatever the time of day, pressing the ON button on the TV control turns the TV on. When integrating smart-home interfaces in an attempt to establish them as new discriminative stimuli for operants requiring electric energy (this type of behavioral flexibility would be probably best captured by the reversal learning paradigm as it concerns switching between different discriminative stimuli), they signal in the current setup of the system only changes in the consequential outcome money loss (e.g., time-of-use tariffs), which is only a subtle change in a consequence outcome not linked to the function of the operant. Thus, when talking about behavioral flexibility in the context of a building / smart-grid system, mostly it addresses questions of behavioral flexibility coming closest to the first connotation mentioned by Bond et al. [150] of (rapid) modification of behavior due to subtle changes in consequences.

In our building / smart-grid setup, we model on the user level the situation of a fixed context. The question is hence not one of user flexibility as discussed in the conceptualizations above, because we do not investigate the effects of changing consequences (or exploratory or playful behavior). Instead, for current living situations, we want to consider the potential to shift user behavior to off-peak times based on the behavioral effort such timely change would require. This does belong to evaluations of user flexibility, but as was detailed above, is not a question of behavioral flexibility in a theoretical sense (independent of connotations), because there is no changing environment. In our study, we describe the *effort* for shifting behavior in time, which is assumed optimally adapted to the current context of a user. As we monetarize the estimation of effort for shifting behavior in time to describe user flexibility, we call this indicator *behavioral adaptive cost* (bac). We integrate bac in the building model to account for barriers and potentials of user flexibility under current context restrictions, as those barriers and potentials will also affect the achievable changes in user behavior for example by setting pricing consequences.

In short, with the previous results from the cluster analysis, which point toward the importance of contextual factors for determining the distribution and variability of behavior, bac are an indicator of the effort for changing the time of a beginning behavior away from the preferred time point, which is assumed to be the optimal time point under a given context restriction. The question is thus, how can the functional relationship between behavioral adaptive costs and varying time differences between the preferred time of use of an electrical appliance be described for every hour within a day?

As a method, we chose a correlational design, which was conducted as an online survey from 3 April until 10 May in 2018 (a date, not sample size was employed as criteria for ending the survey). Predictors are the behavioral activity pattern (weekday (1 through 3) and weekend day (1 through 6)) and type of appliance (washing machine, dryer, dishwasher, stove, coffee machine, TV and computer). To keep the time for the survey agreeable for participants, we limited the appliance types to those with a relatively high impact for which a direct user interaction is assumed in the smart-home and selected from different groups of activities, like doing laundry, cleaning, physical recreation, preparing meals, watching TV and using the computer. Further, participants were randomly assigned to choose either a matching weekday activity profile (which profile of activities matches your weekday activities best?) or weekend activity profile. The chosen profile constituting the fixed context for that participant. Criteria are the behavioral adaptive costs and the preferred time of using an electrical appliance. Behavioral adaptive costs are asked for in Euro on a scale from 0€ to 10€ in increments of 10 Cents for the minimal amount necessary to shift the appliance use behavior away from the preferred time of use for each hour within 24 hours. The preferred time of use for an electrical appliance is asked for in full hours.

In total, 107 people participated. They assigned themselves to the activity patterns (context restrictions) as can be seen in Figure 35. Overall, the evaluated amount of matching seems reasonable (compare Figure 36). In contrast to the number of persons assigned to the different behavioral activity patterns in the cluster analysis, the distribution in this sample differs. During weekdays, most people chose the occupational and educational activity pattern, while in the cluster analysis most persons are sorted in the occupational cluster one and the absence of main structuring activity cluster three. During weekend days, most people selected the activity patterns hobbies, social activities during the day, evening and late-night social activities and occupational activities, while the cluster sizes are largest for both watching TV clusters and social activities during the day.

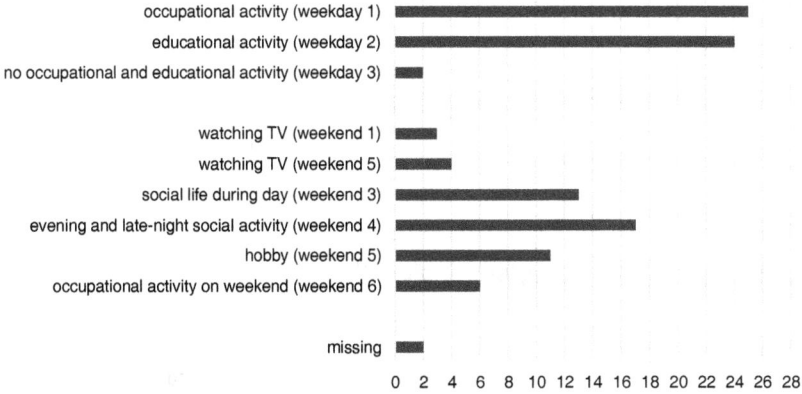

Figure 35: Number of participants per behavioral activity pattern

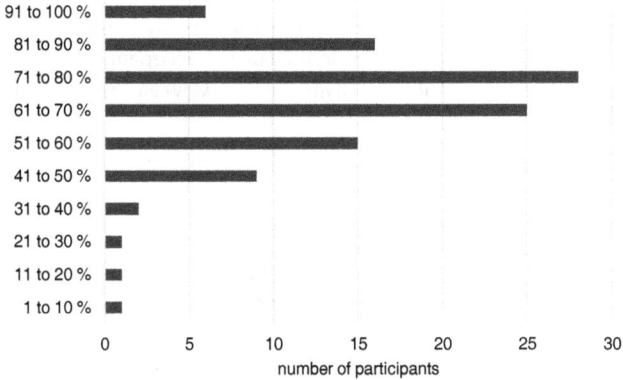

Figure 36: Overall match with behavioral activity patterns

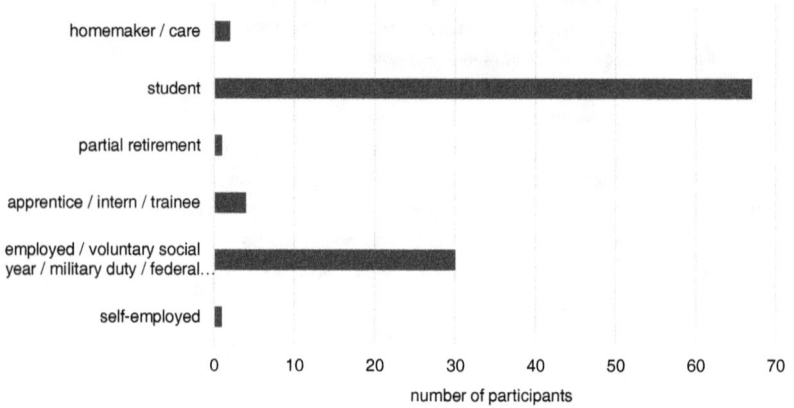

Figure 37: Occupational status

This distribution of activity patterns might be because many of the participants were students and employees as can be seen in the current occupational status Figure 37. For the distribution of gender across activity patterns view Figure 38. Participants are between 18 and 65 years old.

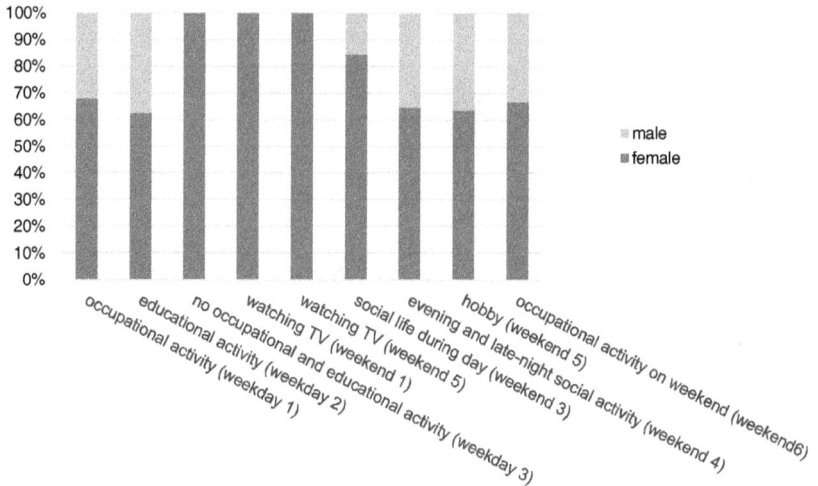

Figure 38: Gender distribution in the behavioral activity patterns

The main outcome for the user-model within the building simulation is the description of the behavioral adaptive costs differentiated for the different clusters and appliance types. First, we analyzed their functional relationship graphically by plotting for each subject the bac on the y-axis for each hour difference from the preferred time of use. As an example, view the Figure 39 for subjects VP_24 and VP_83, who sorted themselves as belonging to weekday cluster 2 and answering the question of effort for shifting behavior away from their respective times of preferred use (0 on the x-axis) for the appliance washing machine.

Figure 39: Plotted behavioral adaptive cost raw data from two participants as examples of two peak versus one peak steep curve type categorization

We then identified similar shapes in the curves and summarized the information on the functional relationship between bac and time shift by aggregating the raw bac values of those similar curves. This meant, having for each appliance type and cluster up to five aggregated bac curves. The curve types are categorized qualitatively as one peak versus two peaks indicating one preferred use point versus two preferred use points, while the "second" preference point can also be associated with higher bac values than the chosen preference point from a participant. An example of this differentiation is VP_24 and VP_83 in Figure 39. One peak curves are further categorized into one peak steep (steep slope around preferred use point, an example is VP_83), one peak fat (less steep slope), one peak flat (several 0 or close to 0 bac values around the preferred use time) and linear. For examples of those curve types view Figure 40 from weekday cluster 2 for the appliance type computer. The shown curve type examples represent prototypes of the chosen categories for summarizing the data, but many categorization decisions were less clear and include simplifications of shape types.

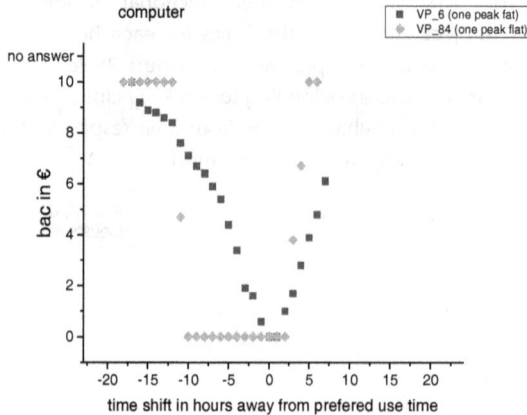

Figure 40: Examples of curve types one peak fat and one peak flat

The raw data in the categorized curve types were then aggregated by averaging the available bac values for each hourly time shift. To pass on this information for the user model in the building would have been sufficient, but for future use and better manipulation possibilities, we fitted two types of functions to the aggregated curve types in order to be able to change parameters in the simulation. Apart from the linear function, we performed a Multiple Peak Fit Analysis with the program *OriginPro* to describe the aggregated bac curves. As can be seen in the plotted bac curve of VP_6 in Figure 40, for some cases a quadratic function would have been an adequate description of a bac curve, especially as it seems to represent the development of bac values around the preferred time of use well. Interesting features of the bac curves, such as multiple peaks indicating several differential preferred use times and upper limits indicating possible restrictions would not have been describable, which is why we chose to describe the bac curves with an amplitude version of the Gaussian peak function with the following form:

$$y = y_0 + Ae^{-\frac{(x-x_c)^2}{2w^2}} \qquad (9\text{-}1)$$

with the parameters y_0 denoting the offset of the curve on the y-axis, A the amplitude, x_c the center of amplitude on the x-axis and w half the width of the amplitude.

Overall, the curve fit was acceptable with an adjusted $R^2 = 0.89$. One curve fit was bad with an adjusted $R^2 = 0.46$ for the appliance TV in weekday cluster 4 for the curve type one peak flat. The derived functions and parameter values describe behavioral adaptive costs in relation to hourly differences from the preferred use time of seven electrical household appliances. The bac values cannot be interpreted in terms of their absolute money values as they are just used for scaling purposes to indicate behavioral effort for shifting electrical appliance use behavior in time. As bac are linked, via study design, to the contextual restrictions identified for the three weekday activity patterns and six weekend activity patterns, the behavioral adaptive costs indicate the behavioral effort of shifting user behavior away from an optimal adopted time point to other time points during a day. Even though an upper limit of bac is inflicted by the question format limiting the scale to 10€ per hour or the *no answer* option, we nonetheless argue, that modeling an upper limit of shifting possibilities is warranted as there are times under the current living situations where people have no possibility of performing a behavior (due to context restrictions). In the open answers of the survey, this showed for example by comments stating that during weekdays there is little flexibility in shifting chores due to work, commute and evening volunteer work (VP_62). From a behavioral psychology point of view, it would be necessary to analyze in future work, in how far lifting or reducing context restrictions would increase the potential of residential user flexibility. After all, in an energy system with substantive changes in the availability and generation of its resource, a perspective limited to analyzing the potential of user behavior changes within a system optimized for different resource characteristics might be too narrow for a successful transformation.

In order to investigate the potential of user flexibility within the building and smart-grid model, the parameter values of the bac curves of the user-behavior model are integrated into the user database from Section 9.2.3.

9.2   Building Model

C. Reinhold

The investigation of the time-related electrical and thermal behavior of buildings in a power supply system requires the modeling of the relevant electrical and thermal devices within buildings and their aggregation at the grid level. As a further component, the user is considered with his behavior modeling. A flexible coupling enables the holistic investigation of building structures with the interaction of user and device.

In the following sections, the building model and the simulation environment elenia Simulation Environment (eSE) will be presented in detail. eSE enables the holistic investigation of scaled problems for connected electrical and thermal devices and systems. The individual elements of eSE are organized in modules so that only a few modules from the simulation environment were used in these investigations. Section 9.2.1 describes the modular simulation platform eSE. The following Sections 9.2.2-9.2.7 additionally addresses the used modules and their structure.

9.2.1 Simulation Environment

The MATLAB-based modular simulation environment eSE [151] has been developed for the variable mapping of energy conditions in buildings. Through the dynamic linking of electrical and thermal systems, as well as control systems, it is possible to investigate highly scaled problems in a single environment. The simulation environment is subdivided into a number of software modules, which can be used individually as well as in coupled investigations. Figure 41 shows the general structure of the simulation platform with the individual modules.

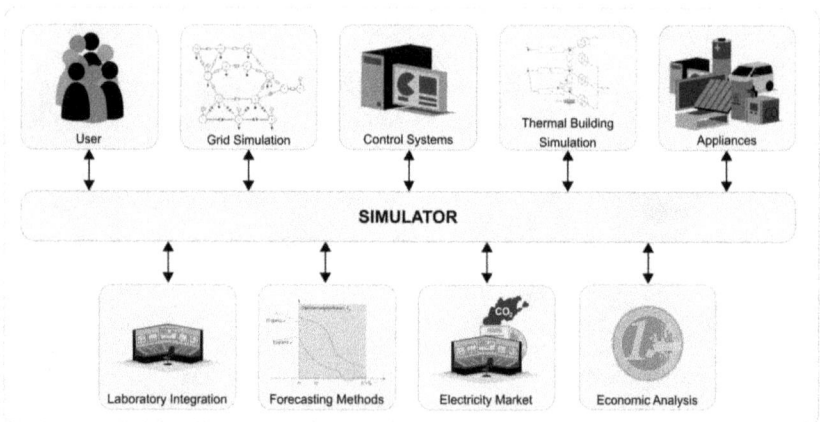

Figure 41: General structure of simulation environment eSE

The core of the simulation environment is the main simulator, which is represented by a MATLAB class. It has a connection to all modules and handles their data management. In addition, the simulator organizes the coordination of information flows between coupled models and stores continuously all relevant information generated during a simulation. For these questions, the modules *User, Control Systems, Appliances, Forecasting Methods, and Economic Analysis* were used and developed, which are described in detail in the following subsections.

9.2.2 Appliance

For computer-based modeling of electrical and thermal devices, two general approaches can be identified in the literature. Statistical, data-analytical models are based on a large database of measured power profiles and other physical parameters of the respective devices. They aim at describing the behavior of devices based on their progressions and to relate them to selected parameters such as season, temperature, global radiation, and household size. Using time series analysis methods such as linear regression or the autoregressive moving average (ARMA models), statistical models can be determined to describe the time series. These models enable data-based reproducibility of the performance profiles and identification of the main influencing factors. Disadvantages of the statistical models are the necessary extensive database and the lack of investigation possibilities for the user and device behavior. [152]

For these reasons, a bottom-up approach was chosen which allows the modeling of individual devices at any level of detail in terms of energy and information technology in order to dynamically generate electrical power profiles depending on the device and connect them with the behavior of users. In general, the available bottom-up models differ in three central aspects:

- Degree of detail of user modeling and database used
- Degree of detail of the database for modeling and generation of electrical profiles of the devices
- Stochastic approach used

The developed models in eSE improve the approaches of the literature by combining their strengths and extending the methods used. The models in eSE have the following basic characteristics.

- Free parameterization of the device models without using a database consisting of measured device load profiles
- Dynamic dependency on the respective weekday or type days.
- Any time resolution of the individual models, as there is no direct dependency on measured device profiles.

- Detailed and parameterizable modeling of the user on the basis of an empirical data set of the time use survey in combination with empirical or stochastic synthetic profile generation

All electrical and thermal models are assigned to the *Appliance* module and are listed in Table 11 below.

Table 11: Modeled Appliances and Control Systems

Appliances		
washing machine	photovoltaic inverter	battery inverter
tumble dryer	photovoltaic module	storage cells
fridge	charging station	energy management system
freezer	electric vehicle	coffee machine
oven	iron	computer
dishwasher	electric stove	hairdryer
heat pump	hifi system	microwave
television	electric kettle	

As an example of the implementation of the *Appliance* module, the class diagram and the example simulation for a refrigerator in Figure 42 are shown. The left side represents the class diagram and the separation into attributes and methods. In eSE, the extended attributes are split into two groups. The dynamic attributes, where the values change with each simulation step, and the static attributes, which are set before the simulation. In addition, the dynamic attributes are stored by the simulator in an HDF5 file for subsequent analysis.

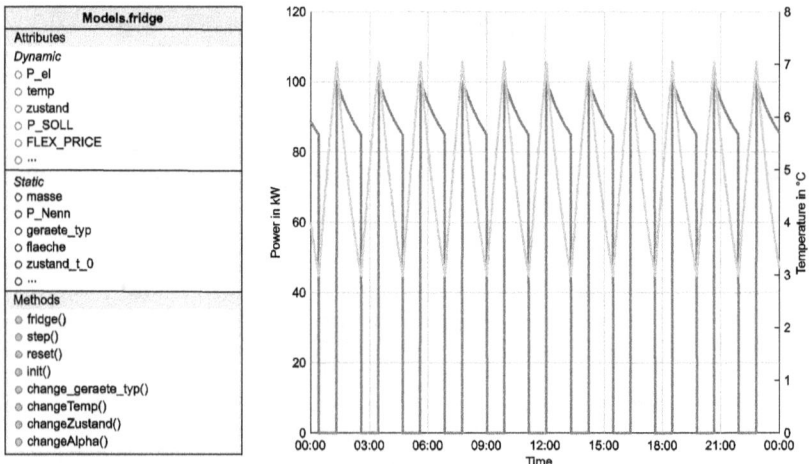

Figure 42: Class diagram (left) and time series of physical parameters (right) of a fridge

The model-independent methods *step()*, *reset()* and *init()* are equal for all models of the module *Appliance* and are called by the simulator if required and are used as a communication interface. The other methods are model-specific and are called within the model-independent methods. The *step()* method executes the calculation steps for the model. The calculated values of the dynamic attributes are queried from the model after each simulation step and stored in the mentioned HDF5 file. The right side additionally shows the resulting time curves of the active consumption power *P_el* and the refrigerator interior temperature *temp*.

9.2.3 User

This Section describes the integration of the user-behavior model from 9.1 into the module structure of eSE and the content of the *User* module. In the first part, the structure of the module is described and the methodical procedure for the technical coupling between the user behavior and the device behavior is shown. The flexible coupling between user and device enables the generation of activity profiles for any user using descriptive parameters. Finally, the integration and parameterization of the empirically determined behavioral adaptation costs from Section 9.1.3 in eSE are described.

Structure

The module user contains several program components, algorithms and data structures, which are described in detail in Figure 43 below. The described user-behavior model from Section 9.1 is extended by the coupling of activity and appliance, the determination of descriptive parameters such as duration of use, frequency of use and the time-related probability of use, and the integration of behavioral adaptation costs into the model. As an interface to the modeling exists parallel to the user-behavior model a user model, which represents the behavior of users in model form.

All information and data records from the user-behavior model are bundled in the form of a local SQLite database with an associated frontend so that access via SQL queries during the simulation and from other programs is guaranteed. The user database contains the datasets from the time use study, the behavioral activity pattern, the coupling of device and activity, descriptive parameters and the behavioral adaptive costs and thus represents the user behavior model. The user model, on the other hand, queries the information from the user database in order to generate activity, device activity and appearance profiles related to the simulation data and the parameterized user with the help of a self-developed algorithm.

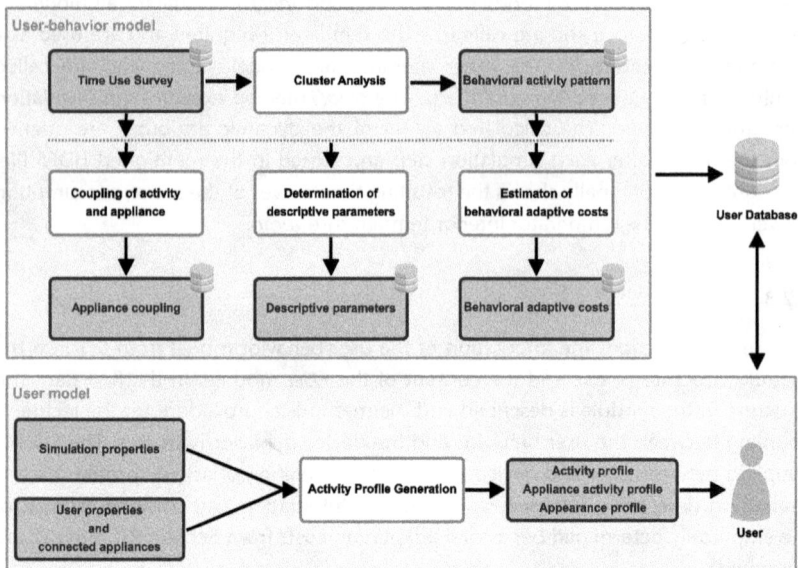

Figure 43: Structure of module User

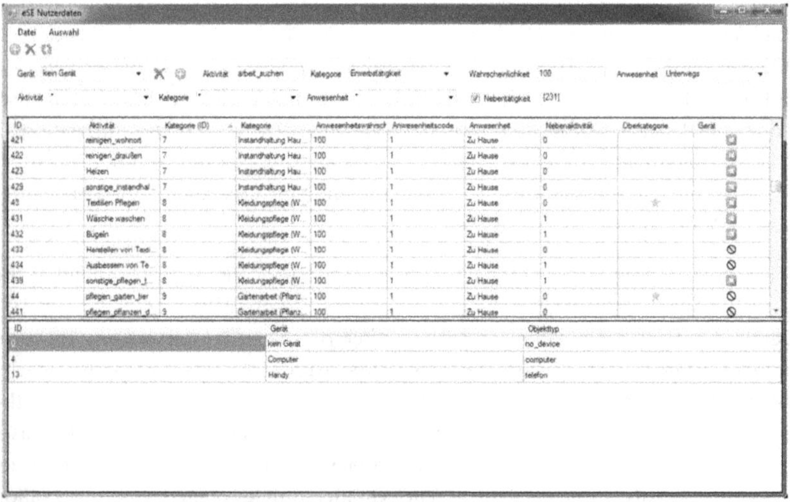

Figure 44: Exemplary view of the user database

Coupling of activity and appliance

The user-behavior model describes activity patterns for people throughout a day. The categorized activities include some for which an interaction with electrical appliances is assumed. This offers the possibility to couple activities, for which an interaction with an electrical appliance is assumed, with electrical load behavior of appliances and arrives at a description of electrical power profiles in residential buildings.

From all the appliances modeled within the building system, direct interaction with a user was assumed for the following appliances: washing machine, tumble dryer, dishwasher, oven, electric stove, iron, microwave, electric kettle, coffee machine, hairdryer, hifi-system, computer, and television. For the other appliances control over the time of use or energy consumption of appliances is managed by the building control system.

The categorized activities from Table 9 are linked to user-dependent appliances. Activities without user interaction with technical consumer appliances are set to default value during linkage. The assignments are entered in the user database in the form of a lookup table. A multiple assignments of consumer appliances to an activity (activity: washing ←appliance: washing machine, tumble dryer) is also taken into account in the database.

Table 12: Lookup table activity (upper category) – appliances [153]

Appliance	Activity	Appliance	Activity
no appliance	sleeping	computer	childcare at home
electric kettle, coffee machine, hairdryer	physiological recreation	*no appliance*	care and support of adult household members
computer	occupational activities	*no appliance*	housekeeping
computer	education	*no appliance*	volunteer work
computer	other education	computer	social activities
oven, microwave, coffee machine, electric stove, dish washer, electric kettle	kitchen work	computer	hobbies, sports, game playing
no appliance	chores	computer	reading
washing machine, tumble dryer, iron	clothing care	television	watching TV
no appliance	gardening and animal care	hifi system	listening to radio and music
no appliance	handicraft activities	computer	using a computer or smartphone
no appliance	shopping	electric vehicle	travel and commute activities

The appliance activity coupling and the user database can then be used to determine three descriptive parameters (time-related probability of use, duration of use and frequency of use per day) for each appliance and each activity. These parameters form the basis for the synthetic procedure for generating activity profiles.

Determination of descriptive parameters

Based on the data from the time use survey and the behavioral activity patterns, the descriptive parameters of the time-related probability of use, duration of use and frequency of use of all activities can be determined. Using the lookup table between activity and appliances, the descriptive parameters of the appliances are calculated. The following sections show the results of these three parameters.

Time-related probability of use

The time-related probability of use of activity or appliance reflects the probability that an action will occur at a given point in time. It is synonymous with a probability function that assigns a probability value p in the interval $[0,1]$ to a random number T, in this case, the time t. The probability value p is the time of the action.

$$p = f(t) = P(T = t) \tag{9-2}$$

The time-related probability of use is determined by selecting the time use survey data by weekday and cluster. The absolute and relative frequencies of the activities are then determined for each point in time. This results in a probability specification per activity for each point in time. The activity/appliance lookup table was used to define the time-related probability of use for each appliance.

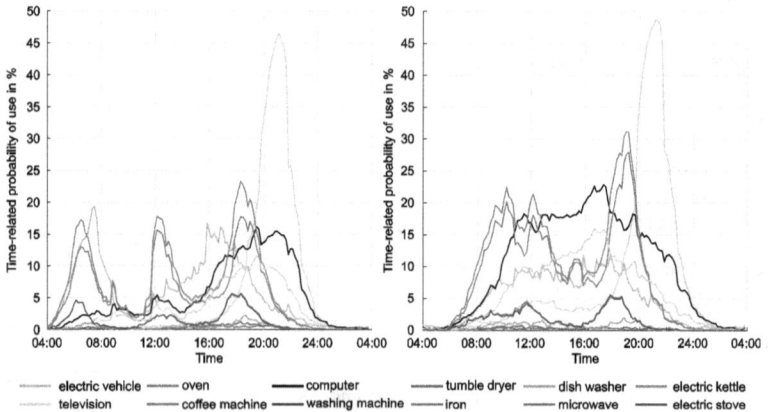

Figure 45: Time-related probability of use for appliances (left: weekday Cluster 1; right weekend Cluster 3) [153]

Figure 45 shows an example of the time-related probability of use of the considered appliances for one weekday of cluster 1 (left) and one weekend of cluster 3. The high probability (>45%) that a television set will be used in the evening hours is striking. The time-related probability of use was calculated for all activities and appliances depending on the day type and the cluster.

Duration of use

The duration of action indicates how long the action is executed continuously. To calculate the duration of use, the time use survey data is filtered and the start/end time and duration of each activity entry are determined. Each duration of the use data point is directly linked to an activity and a time use survey data point. Using the lookup table of activity/appliance, the data points are linked to the appliance. Thus, activities and appliance have any number of duration of use data points. From these data sets, suitable probability functions are calculated using an automatic MATLAB fitting algorithm. These functions are used to calculate the stochastically distributed duration of use values using a synthetic method. Table 13 lists the mean value, standard deviation and probability function of weekday cluster 1.

Table 13: Duration of use for the appliances (weekday cluster 1) [153]

Appliance	Mean value in minutes	Standard deviation in minutes	Distribution function in seconds
television	84.701	57.532	Gamma ($a = 2.126; b = 2390.172$)
electric vehicle	26.147	24.873	Generalized Extreme Value ($k = 4.328; \sigma = 34.599; \mu = 607.988$)
oven	22.005	16.653	Generalized Extreme Value ($k = 4.6995; \sigma = 33.996; \mu = 607.23$)
hifi system	36.298	32.713	Inverse Gaussian. ($\mu = 2177.9; \lambda = 2959.315$)
computer	35.707	34.541	Generalized Extreme Value ($k = 0.6920; \sigma = 684.743; \mu = 1082.7$)
electric stove	21.679	15.770	Generalized Extreme Value ($k = 4.773; \sigma = 51.982; \mu = 610.8792$)
iron	42.566	27.897	Birnbaum-Saunders ($\beta = 2008.421; \gamma = 0.733$)

Frequency of use

In addition to the duration of use and the time-related probability of use, the decisive factor for action is how often executed per day. This property describes the frequency of use. Equivalent to the calculation of the duration of use can be proceeded also with this parameter, only in this case the number of repetitions per diary record is determined. If the activity is not performed in this data record, the amount of 0 is used in the evaluation. After linking the activity and appliances, the probability functions for the usage frequency are also determined using the same methodology Table 14 shows the stochastic characteristic values for the individual devices.

Table 14: Frequency of use for the appliances (weekday cluster 1) [153]

Appliance	Mean value	Standard deviation	Distribution function
coffee machine	2.659	1.118	t- location-scale ($\mu = 2.647; \sigma = 1.059; \nu = 19.339$)
washing machine, tumble dryer	0.234	0.608	Generalized Pareto ($k = 7.434; \sigma = 0; \theta = 0$)
electric kettle	3.502	1.593	Generalized Extreme Value ($\mu = -0.113; \sigma = 1.425; k = 2.831$)
dishwasher	0.362	0.713	Generalized Pareto ($k = 10.999; \sigma = 0; \theta = 0$)
television	1.035	0.806	Generalized Pareto ($k = -0.186; \sigma = 1.212; \theta = 0$)
electric vehicle	3.246	1.745	Generalized Extreme Value ($k = -0.077; \sigma = 1.485; \mu = 2.5$)
computer	2.130	1.896	Generalized Pareto ($k = -0.137; \sigma = 2.422; \theta = 0$)
oven	0.864	1.016	Generalized Pareto ($k = 20.492; \sigma = 0; \theta = 0$)
hifi system	0.065	0.279	Generalized Pareto ($k = 3.354; \sigma = 0; \theta = 0$)
electric stove, microwave	0.850	0.996	Generalized Pareto ($k = 20.413; \sigma = 0; \theta = 0$)

Generation of activity profiles

Based on the data of the time use survey and the descriptive parameters, two methods for generating user profiles are available here. Activity profiles, device activity profiles and appearance profiles for the respective selected users are calculated independently of the selected method. The profiles are comparable with the standard load profiles for an unregistered load profile measurement, only that the behavior of users is estimated at this point.

The empirical method uses the time use survey and the previous assignment to the clusters as a data basis. The input variables are the simulation settings (start time, end time and simulation step size), the user characteristics and connected devices. The necessary information (day type, cluster) is determined for each day. This daily information is used to query the associated diary entries of the time use survey from the user database. An entry is randomly selected from the number of diary entries and an attempt is made to add it to the list of the activity profile. Before this, the activity of the last existing entry and the first activity of the current entry are checked for consistency. For example, if the person is not present, but the following activity forces an interaction with an appliance, then a mobility activity is missing to get back home. If this happens, another diary entry will be randomly selected until a valid state is created. The activities of the entry are then connected to the appliance using the lookup table between activity and appliances. In this way, the appliance activity profile is created. Each activity simultaneously implies whether the person is at home or on the move, resulting in an appearance profile. The sequence of diary entries leads to the activity profile. All profiles are assigned to the user. To illustrate the procedure, the individual steps of the process are shown in Figure 46.

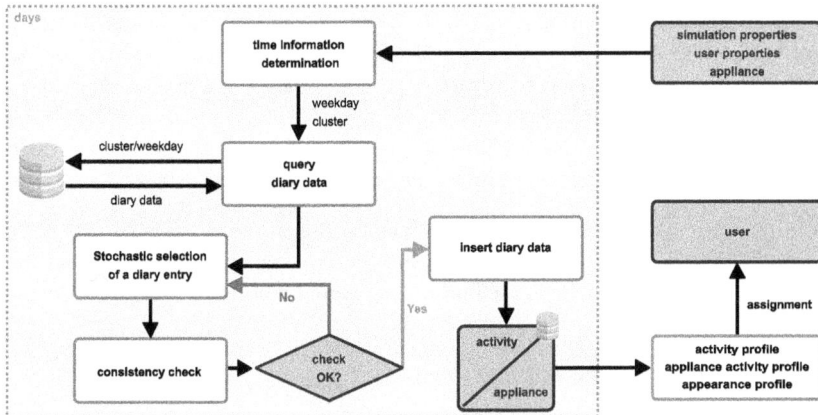

Figure 46: Process representation of the empirical method [153]

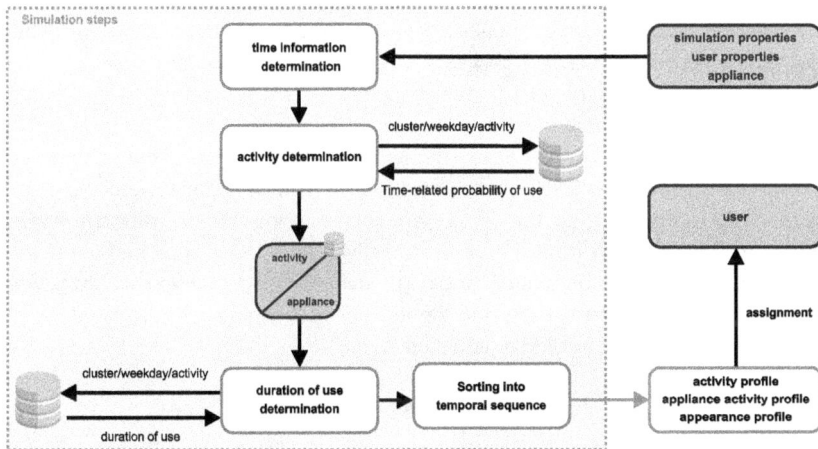

Figure 47: Process representation of the synthetic method [153]

The synthetic method creates an activity, device activity and appearance profile based on the descriptive parameters without using the data of the time use survey from the user database. In this way, it is possible to synthetically simulate behaviors of different groups of users with the help of a small amount of data. Figure 47 shows the process representation of the synthetic method.

The input variables are identical to the empirical method, except that the iteration loop is run through for each simulation step. Thus, for each step, the corresponding time-related information is determined. Activity is then determined by querying the usage probability of all activities from the user database. An activity is selected stochastically from this distribution. The determined activity is then connected to an appliance. After the coupling, the respective activity duration is determined. For this purpose, the necessary data records from the user database are queried again, only that the determined activity is filtered at this point. Finally, the activity or appliance activity is added to the existing time series. After all simulation steps have been completed, the user-dependent profiles are created, which are assigned to the user model in the last process step.

Due to the improved parameterizability and the non-use of the datasets from the time use survey, the synthetic method is used within the investigations and in the further process.

Integration of behavioral adoption costs

The behavioral adaptation costs determined in Section 9.1.3, allocated to the respective behavioral clusters, are stored within the user database and can be specifically queried. During the static coupling of appliances with the users within the dwelling units, the curves are filtered according to the cluster assignment of the day type (weekend, weekday) and assigned to the respective appliance with the unique identification of the user. Furthermore, the determined monetary dimensions of the behavioral adaptation costs are scaled by a factor f of $1/100$ in order to achieve harmonization with other monetary variables within the distributed optimization at the grid level and to avoid a dominant behavior toward the other optimization goals. During the simulation, the assigned curves are filtered with regard to the time shift from the preferred activity time in order to determine the user-specific costs for a behavior change. The behavioral adaptation costs bac_{r-s} of the alternative schedules s, related to the reference schedule r are calculated for all alternative schedules according to equation (9-3).

$$bac_{r-s} = bac_r - bac_s \ \forall s \in S \tag{9-3}$$

Positive values represent an incurred effort for the user. Negative values, on the other hand, mean a time shift of the alternative schedule at a suitable time for the user. If the values are zero, there are no improvements or deteriorations for the user. The determination of possible time shift corridors is explained in Section 9.2.4 device-specifically.

9.2.4 Power Flexibility

A directed influencing of the time-related electrical behavior and the modification of stationary appliance characteristics as a reaction to an external signal with the aim of providing a service in the energy system can be described as flexibility. In the present investigations, the active power of the appliance is shifted in time and varied in height. For self-learning, schedule-based and grid-side optimization strategies and for estimating the appliance behavior for a variable time range, it is necessary to generate any number of feasible alternative schedules in addition to the reference schedule. The reference schedule represents the time-related active power behavior under the influence of the building-based control system. The model technical illustration of the alternative schedules is realized accordingly by methods of sampling. Within the framework of the investigations, device-type-specific schedule generation algorithms are used, which are derived from the reference schedule and the device properties and calculate alternative schedules. Except for the storage systems of the buildings, 30 feasible schedules are set for the sampling methods. The algorithms developed in each case are described textually in the following sections and their effects on schedules are shown.

User-driven appliance

A schedule generation algorithm is only available for appliances with empirically determined behavioral adaptation costs. In this investigation, the appliances are washing machine, dryer, dishwasher, coffee machine, television, electric stove, oven, and computer. Accordingly, the other appliances are not made more flexible and have only a reference schedule and no alternative schedules.

The first step is to check whether the user used the appliance during the simulation time. If this is not the case, no shifting potential is determined and the alternative schedules correspond to the reference schedule. Afterward, alternative start times are determined for all switching operations of the appliance, so that there is a time shift in the use of the appliance. The restriction for all appliances is that the start time must lie within the simulation time. For the appliances coffee machine, television, electric stove, oven and computer it is additionally defined that the usage time can only take place during the appearance of the users and that the use of the appliance must be finished before the activity sleep or the next absence phase. The washing machine, dryer, and dishwasher can also be used without the presence of the user. If no free shift time is determined, the start time of the reference timetable is used. For the washing machine, the active electrical power of the reference schedule and the next two alternative schedules with the sleep phase and the presence phase of the user is shown below.

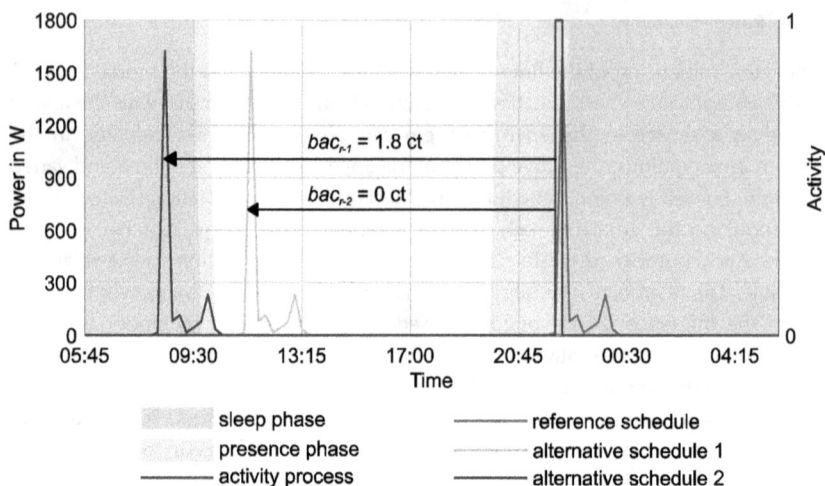

Figure 48: Power flexibility for a washing machine

In this example, the washing machine is switched on once and shifted within the valid restrictions. Each variation of the reference schedule causes a change in the user's behavior, which in turn is reflected in the relative behavior adaptation costs bac_{r-s}. In the example shown here, alternative schedule 1 has a positive value of $bac_1 = 1.8\ ct$ and alternative schedule 2 has a value of $bac_1 = 0\ ct$.

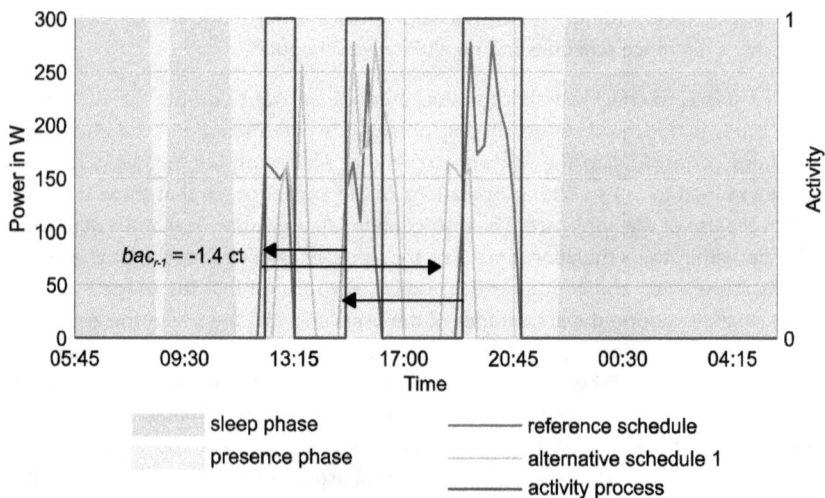

Figure 49: Power flexibility for a computer

Especially user-dependent appliances with low power are switched on and off several times during the day. Figure 49 accordingly illustrates the situation for a computer and its temporal displacement possibilities. The relative behavioral adaptation costs for the alternative schedule bac_{r-1} are the sum of the shifts of all switching operations. In this example, the activities are shifted to a suitable time period for the user.

Charging station with electric vehicle

Electric vehicles have an integrated storage system and are charged via charging stations in the public and private sectors. Charging at public charging stations is excluded in this investigation. During driving, the storage system discharges according to the underlying driving profile of the user. The electric vehicle is electrically and communicationally connected to the charging station before the start of the charging process and can charge the storage system until the start of the next driving cycle. Technically, the charging power can be varied in the range $[0; P_{N,LS}]$ up to the maximum power level of the charging station $P_{N,LS}$. Bidirectional charging is excluded at this point. The objective for the reference schedule and the alternative schedules is to achieve the maximum energy capacity of the storage system until the next driving cycle begins. The next departure time is known to the control system at any time based on the activity profiles of the user.

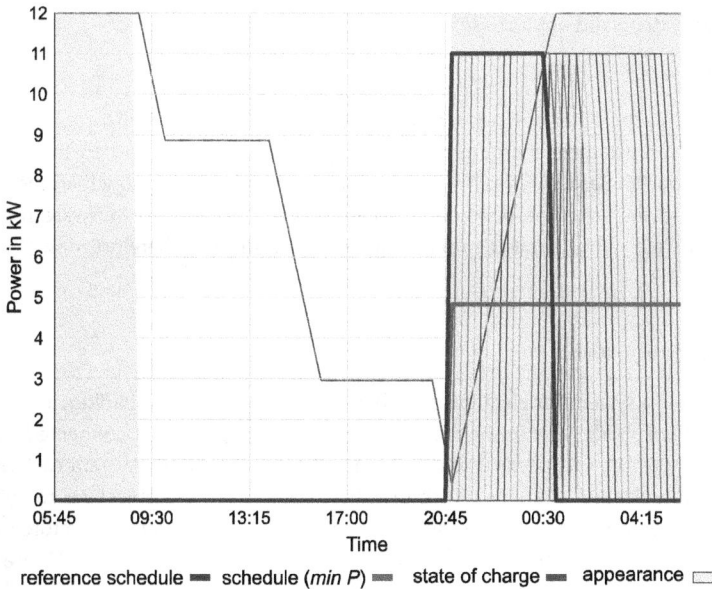

Figure 50: Power flexibility for charging station and electric vehicle

121

The reference schedule represents the loading strategy of immediate loading without considering other influencing factors. With the alternative schedules, the start times of the charging process are gradually shifted in time to the size of the simulation step, so that the storage system is fully charged at the end of the charging process. In addition, the last alternative schedule implements a charging strategy with a maximum possible limitation of the charging capacity (min_P), so that the minimum load on the grid system results, see Figure 50. For all alternative schedules, the relative behavioral adaptation costs bac_{r-s} are assigned the value 0, since a temporal shift of the charge profile due to the restriction of the full charge level has no direct influence on the behavior of the users.

Photovoltaic system

The inverters used for photovoltaic applications are able to flexibly vary the active power at the output in the second's range in relation to the current direct current power of the photovoltaic modules. An increase of the generation power can only be realized with an appropriate provision, i.e., with a static reduction of the generation power, which proves to be uneconomical in current applications and under consideration of the current regulatory conditions. A reduction of the generation capacity is nowadays carried out for reasons of grid and system stability, which in turn leads to a degradation of the economic efficiency from the point of view of the owner of the device. For these reasons, the coupled analysis of building and grid systems did not include any flexibilization of PV systems.

Storage system

In contrast to the other devices, storage systems have the characteristic of being able to flexibly adjust their charging and discharging power according to the current energy capacity. Based on this, this device is assigned a higher priority than the other devices and the schedule generation is carried out in a separate software module outside of eSE. The detailed implementation procedure is therefore described in Section 9.3.3.

9.2.5 Control Systems

As the control system for electrical and thermal systems within a building, an energy management model acts in the simulation environment, which is connected to the respective devices via specific exchange parameters. Each control system can have a specific optimization strategy, which uses the strategy and the objectives to control the connected devices with the required target values. The sequence structure of the signal exchange between the control systems and the devices is shown in Figure 51.

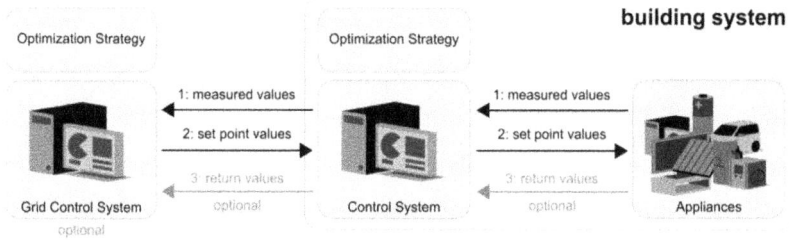

Figure 51: Design and Control Structure in Building Systems

The control system of the building can optionally communicate with a grid control system to be integrated into a plant cluster. In the first step (1), the devices transmit their measured values to the building control system, which in turn transmits aggregated values to the grid control system. The setpoints (2) for the devices or building systems are then calculated based on the optimization strategy and sent to the controllable devices.

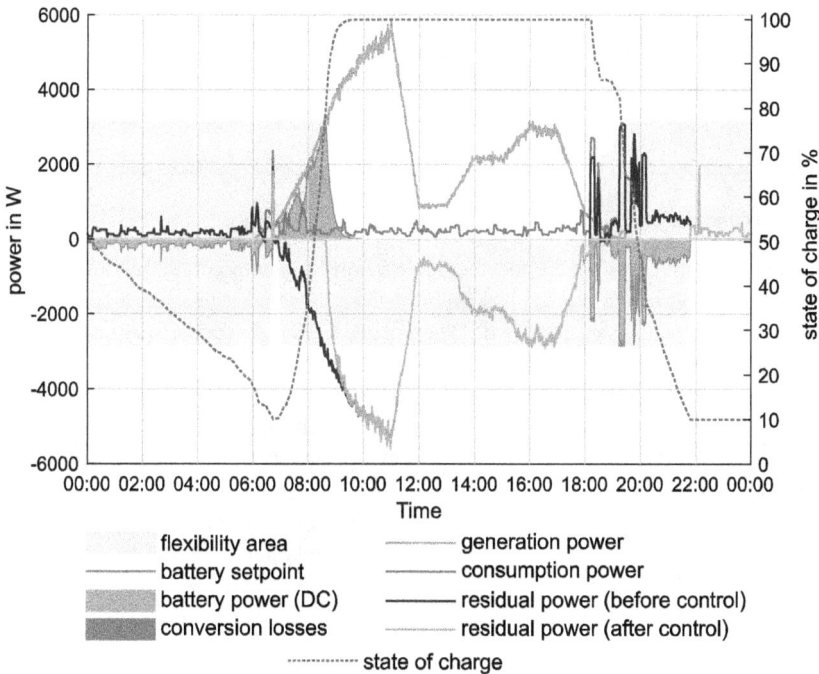

Figure 52: Exemplary power flows and information signals in a building system with a photovoltaic and storage system

The devices adapt their behavior to their technical restrictions. As an option, the implemented setpoints (3) can be sent back to the control systems in the form of feedback control in order to carry out a setpoint comparison. Figure 52 shows the resulting power and information flows of a building energy management system for a controlled photovoltaic storage system to illustrate the control process that is described.

9.2.6 Forecasting Methods

For the implementation of schedule-based operation strategies on superior control levels, the prediction of the time-related appliance behavior is a necessary element. This implies in the previous step the forecast of external data sets for a variable time range in the future, such as global radiation, outside temperature, and wind speed. Within the models, these data sets are used together with static properties and other data sets from other models to predict the appliance behavior and the energetic time-related curves. For the forecasting of the values and the profiles, a number of different methods are implemented, which are bundled in the module *Forecasting Methods*. In the context of the investigations an ideal forecast with a horizon of 24 hours was used, which means 96 single values with a simulation step size of 15 minutes, in order to suppress possible influences by forecast errors and deviations within the control and optimization algorithm.

9.2.7 Economic Analysis

For the economic evaluation and the comparison of system configurations among each other, the economic characteristics of the devices and the associated device operators and actors are modeled. Different evaluation methods, such as the net present value method or the annuity method, are integrated into the models of the devices. Thus, it is possible to investigate the technical behavior of the device and the economic effects in a framework. Each actor can be linked to any number of devices, which represents an assignment of ownership. As a rule, the economic indicators are calculated for all devices of the actor and aggregated at the actor level. In addition, it is possible to develop and integrate actor-specific calculation algorithms. This structural structure results in close integration of the *Appliance* and *Economic Analysis* modules.

9.3 Smart Grid Model

M. Nebel-Wenner

In the following section, a smart grid model is introduced, which simulates and analyzes the potential of cooperative load scheduling of smart buildings. In Section 9.3.1, the topic is introduced and the potential benefits of load management for smart buildings are highlighted. Section 9.3.2 then describes the multi-agent system ISAAC and the heuristic COHDA, which were used for several simulations. In Section 9.3.3 we shed light on the calculation and representation of flexibility in the model. Section 9.3.4 then describes the multiple optimization goals, which were pursued in the model. Details regarding the coupling of the smart grid model with the building model are given in Section 9.4, while the simulation scenarios are described in Section 9.5. Section 10.1 will later present the results of the simulations.

9.3.1 Load Management of Smart Buildings

Flexibility in production and consumption of electrical power within smart buildings may be a key factor in an energy system based on solar and wind energy [154]. In this context, an intelligent and efficient operating strategy enables to optimize the flexibility of various buildings regarding the problem of imbalance of generation and consumption as well as regarding the power supply from the superposed grid. In that respect, a smart grid operating strategy on the low voltage level may the need for grid expansion in the higher voltage levels. In this project, a smart grid model simulates and analyzes the potential of cooperative load scheduling of smart buildings. The input for the smart grid model is received from the smart building model, in which flexibilities of various smart buildings in a low voltage grid are computed (see section 9.2). The results of this model constitute typical load curves in a smart low voltage grid and serve as input for the optimized distribution grid planning model (section 9.7). Additionally, an evaluation of the smart grid operation strategy regarding sustainability has been performed (see section 0)

Virtual power plants (VPPs) are a concept of pooling numerous distributed and renewable energy resources (DER), which also encompasses flexible loads of buildings. Aggregation is necessary to overcome market barriers and for feasible risk management. A VPP usually relies upon a software system to remotely and automatically optimize generation or demand [155]. For optimal use of the unit's flexibility, the scheduling of operation times is required.

Multi-agent systems (MAS) and distributed algorithms have received increased attention in the context of smart grid research, especially within the simulation of VPPs [156]. Multi-agent Systems contain an environment, objects and multiple intelligent agents that solve a common problem (see [157] for further information on MAS). Following the definition of fundamental agent theory in [158], an intelligent agent is defined as a computer system that has different properties, such as autonomy (agents operate without the intervention of humans or others), social ability (agents interact with other agents), reactivity (agents perceive their environment) and pro-activeness (agents exhibit goal-directed behavior).

The advantage of a distributed algorithm lies in the scalability, whereas the use of agents allows abstracting from the specific characteristics of the single unit. MAS can be used as distributed control including a huge number of units, e.g., for scheduling for market optimization or grid stability.

To investigate the flexibility potential of buildings regarding different goals, various simulations of an optimization process of flexible loads of buildings in the low voltage grid have been executed in this project. For this, the multi-agent system ISAAC has been used.

9.3.2 Multi-Agent System ISAAC

ISAAC[1] is an energy unit aggregation and planning software, based on aiomas[2], a lightweight MAS framework written in python. Nieße and Tröschel present it in [159]. The main use case of ISAAC is the aggregation of DERs for VPPs. In ISAAC, each agent represents one energy unit, from which it has knowledge about the local information such as the flexibilities. The different agents in the MAS are connected through a small world overlay network. An overlay network describes the connection of units by virtual links that are built on top of the physical connection of the grid. A small world topology describes a graph, in which each node can be reached from every other node in the network, but not all nodes have to be directly connected.

Agents in ISAAC implement a modified version of the COHDA (combinatorial optimization heuristic for distributed agents) algorithm for solving the optimization problem of scheduling the unit's operation times. COHDA is a fully decentralized optimization heuristic that uses self-organizing mechanisms to optimize a common target and is presented by Hinrichs and Sonnenschein in [160]. In COHDA, agents cooperatively negotiate with each other in order to optimize a global objective function. The heuristic exhibits convergence and termination is robust against single communication faults and it shows good scalability. Fast convergence of the COHDA heuristic depends on massively parallel communication. If applied to the real world, long-term- evolution (LTE) standards such as 3G / 4G or DSL are thus recommended as communication technologies [161].

---

[1] https://github.com/mtroeschel/isaac
[2] https://aiomas.readthedocs.io

In order to prevent undesired behavior, ISAAC is embedded into a controller/observer architecture (see Figure 53). In this setting, on top of the interconnected unit agents, an observer agent and a controller agent exist in the MAS. The observer agent constantly monitors the self-organized outcome of the MAS and passes information to the controller agent, if necessary. The controller agent is responsible to set up the MAS by creating an overlay network and receiving and communicating the objective function. It additionally performs control actions to alter the optimization process, e.g., assuring termination of a negotiation within the desired time. Generally, the controller/observer architecture combines the benefits of self-organized system behavior with the possibility of avoiding unwanted behavior [159].

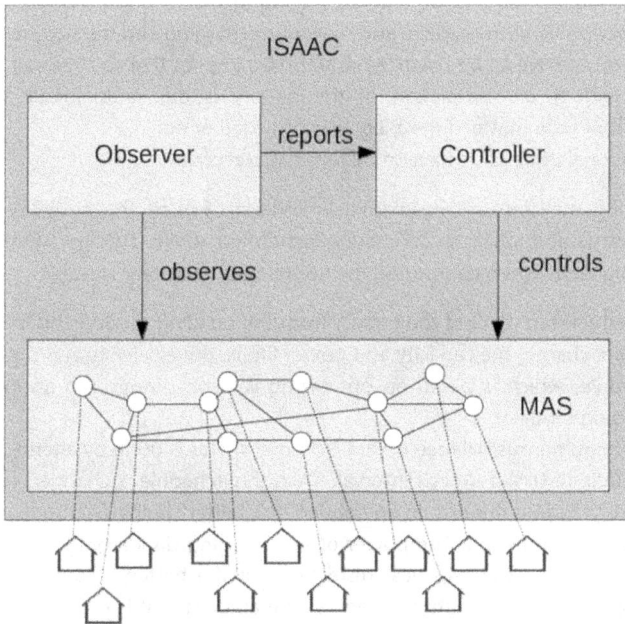

Figure 53: Structure of the multi-agent system ISAAC. Adopted from [159]

9.3.3 Flexibility in ISAAC

In NEDS, flexibility is represented by a number of feasible schedules for each building. Each schedule consists of 96 values, one power value for each of the 15-minute intervals within one day. Within the building model, the flexibility of the different electrical appliances is aggregated and 30 feasible schedules per building are created (see Section 9.2), which are then used in ISAAC. However, the flexibility of battery storage systems is computed apart from that, since the flexibility space of battery storages is generally significantly greater than the one of other electrical appliances in the building.

In order to deal with the large flexibility space of battery systems, we developed a storage agent that is responsible to compute flexibility for battery storages placed at households with a photovoltaic system. There are certain restrictions that have been taken into account when computing alternative schedules:

1. The main use case for battery storages at the building level is the optimization of self-consumption. Therefore, the different feasible schedules must respect this primary use case. This implies, that all surplus power produced by the PV system must be stored in the battery, if possible.
2. All schedules have to be feasible regarding the technical restrictions of the battery (maximum capacity and maximum charging and discharging power)
3. Schedules have to be balance-neutral. This implies that the state of charge at the end of the simulation of the battery is the same in all alternative schedules, no matter if flexibility was provided or not.
4. Additional cycle costs have to be taken into account.

The following method was developed in order to exploit the flexibility space of battery storages and fulfill the conditions mentioned above. Storage agents execute the following steps when computing the flexibility for a battery storage:

1. Get all relevant data of the battery from the building model. This includes the state of charge, the capacity and power limits, the efficiency and the reference schedule, which is based on optimizing self-consumption in an immediate-charging mode.
2. Create numerous balance-neutral schedules. This is done by altering a default schedule of 0kW for each interval. Then this schedule is changed within two distinct random periods. In one period, the battery is charged and in the other one, it is discharged. The power of the charging/discharging signal is varied depending on the technical restrictions of the battery. However, the total amount of energy discharged equals the amount that is charged, taking into account efficiency losses. Figure 54 shows an exemplary set of such schedules.
3. Create a set of potential schedules for each battery by summing up each of the balance neutral schedules created in step (2) with the default schedule received in step (1).
4. Test each of the potential schedules on validity regarding the technical restriction of the battery and regarding the primary use case of optimization of self-consumption.
5. Store each of the valid schedules in the set of possible schedules. For each of them calculate the additional cycle costs on the basis of the investment costs

of the battery divided by the minimum number of cycles that the producers of the battery grant for.

The parameters that have been chosen for the calculation of the schedules are displayed in Table 15.

Table 15: Parameter setting for the computation of balance neutral battery schedules

Possible durations of charging period (min)	Possible durations of the discharging period (min)	Possible energy quantity charged/discharged (% of total capacity)	Number of charging/ discharging periods
{60, 120}	{60, 120}	{10, 20, 30, 40, 50, 60, 70, 80, 90}	1

Given this calculation of flexibility described above, each agent that represents a building may choose among 30 possible schedules while each agent representing a battery storage system has a set of ca. 2000 – 4000 possible schedules, each of the schedules assigned with a cost value.

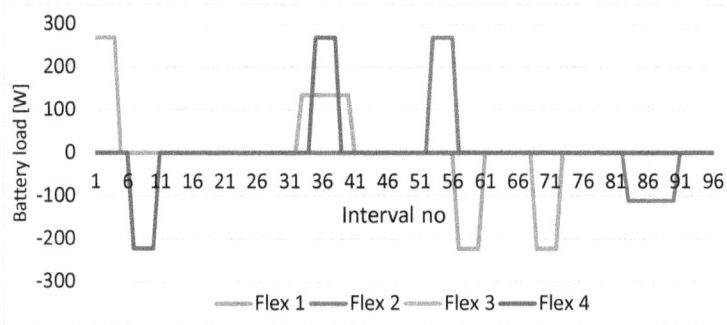

Figure 54: Exemplary balance neutral variations for a battery storage

9.3.4 Optimization Goals

In the NEDS project, the buildings that are part of the VPP optimize their schedules regarding multiple optimization goals. The following high-level optimization goals for the VPP are considered:

1. Optimization of power consumption and power generation
2. Minimization of the peak grid load
3. Minimization of electricity costs at the electricity market
4. Minimization of behavioral adaptation costs

From a mathematical point of view, the set of Pareto optimal solutions constitutes the solution to a multi-objective optimization problem. However, in this setting, one solution must be picked. The most common approach to solve such problems is by scalarization, which then involves formulating a single objective function. This approach is in accordance with the design of COHDA, which only allows a distributed optimization of one single objective function.

Generally, there are several methods to perform multi criteria optimization (see [162] or [163]). Picking one out of the set of Pareto optimal solutions usually involves a decision maker, which expresses preferences on the criteria. This can be done by ordering or weighting the single criteria or by defining additional constraints (e.g., "criteria x must exceed value y"). However, as the objective criteria may have different magnitude, normalization of objectives is required to get a solution that is in accordance with the weights of the decision maker [164].

In the following, the chosen optimization approach will be displayed. In [165] the multi-objective optimization approach is explained and analyzed in further detail with a particular focus on the multi-objective nature of the optimization problem. In order to include the four criteria into one target function, a monetization approach was chosen. The high-level target function describes the costs of an aggregated schedule s of all households that are part of the VPP and consists of four elements:

$$f(s) = \varepsilon(s) - \phi(s) + \gamma(s) + \alpha(s) \tag{9-4}$$

$\varepsilon(s)$ describes the costs at the electricity market, $\phi(s)$ depicts the payments due to feed-in to the grid, $\gamma(s)$ describes the costs of the grid usage and in $\alpha(s)$ the behavioral adaptation costs are computed. In the following, we will explain each of the four sub-functions in detail.

In $\varepsilon(s)$ the costs of the aggregated power consumption schedule at the electricity market are computed. We assume a variable electricity price scheme. The market prices in the simulations are based on historical data from the German Federal Grid Agency (Bundesnetzagentur)[3]. The day-ahead market prices for the years 2015 – 2018 were taken as a basis and aggregated in an hourly resolution for each of the different type of days described in Section 9.5. Figure 55 shows an example of the computed price function for a working day in summer. The computed price functions are constant within all alternatives and simulated years.

In $\phi(s)$, the payments due to-feed in are computed. We assume no subsidizes for feed-in of DERs and hence the feed-in payment is based on calculations at [5]

---

[3] see www.smard.de

regarding the electricity generation cost for 2050 within the scenario of 80% greenhouse gas reduction, which is assumed to be 8.3 cent/kWh.

$\gamma(s)$ describes the fees for the grid usage. The function is oriented on the cost structure of the grid charge in Germany for users with power measurement. The costs are based on two positions: the overall consumed energy within one year (in kWh) and the maximum load within one year (in kW). The prices for each of these positions are set based on the fees that the regional utility EWE charged in the year 2018[4]. However, since we simulate only single days, the fees were scaled down to one single day. In our simulations, we included the grid usage in both directions, meaning that a feed-in peak is treated the same way that a peak of the power taken from the grid is treated.

Finally, $\alpha(s)$ describes the behavioral adaptation costs. They are computed in the building model and assigned to each building schedule (see Section 9.1.3 and 9.2.3). They are summed up for all buildings and remain unchanged within the target function.

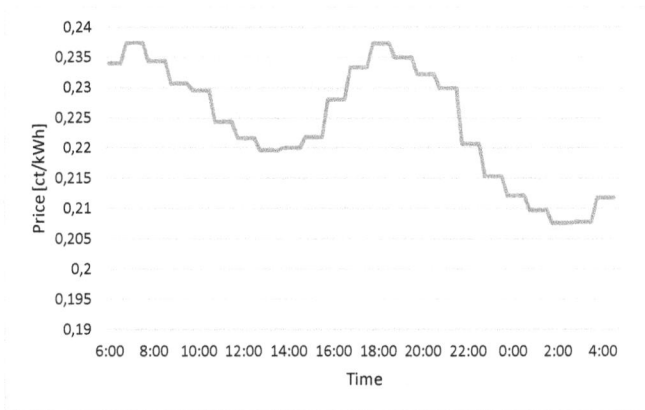

Figure 55: Price function for electricity for a working day in summer computed on the basis of historical data

By using this target function, all high-level optimization goals are included in the optimization process. Since $f(s)$ describes the costs for the VPP, the optimization goal is to minimize this function. It is assumed that all households within the VPP are cooperative and hence try to minimize the costs for the VPP.

---

[4]see https://www.ewe-netz.de/~/media/ewe-netz/downloads/2018_04_03_ewe_netz_nne_strom_2018.pdf

9.4   Coupling of Building and Smart Grid Model

M. Nebel-Wenner, C. Reinhold, J. S. Schwarz

The building model and the multi-agent model ISAAC are directly coupled. As input, ISAAC receives 30 feasible schedules on a 15-minute time resolution and its associated flexibility costs for each residential unit from the building model. Additionally, the reference schedule and technical parameters of the storage systems are sent from the building model to ISAAC. Within ISAAC, the scheduling of the flexible buildings is optimized and finally, the chosen schedules are returned to the building model. In order for each household to represent a separate unit in ISAAC, multi-family houses were divided into several units. Thus, in a 3-family house, 3 residential units and the entire building are separately connected to ISAAC. The entire building is nevertheless calculated in the first step with the consumption and generation data of all appliances from all residential units for the local optimization strategy of the control system. In order to avoid a duplication of the consideration of the individual residential units, the information of the individual residential units was subtracted again from the entire building after completion of the operation strategy, so that only the real power data of the systems at building level were included.

For the coupling of both models, the co-simulation framework mosaik[5] is used [132, 26]. It provides a protocol for the communication between the simulation models, which is called mosaik API (application programming interface). This mosaik API has to be implemented by each simulation model. For the implementation, mosaik provides a low-level API based on TCP/IP and several high-level APIs for the most common programming languages.

ISAAC has an interface to mosaik, which uses the Python-API. For the building model, the mosaik-MATLAB-API was used. The API was enhanced for the simulation in NEDS to optimize the performance. As the building is modeled rich in detail and more than 100 residential units are simulated, the performance plays an important role in the execution of the simulation scenarios. MATLAB offers a toolbox with features for parallelization of code. To achieve reusability for future usage with mosaik, the parallelization was not implemented directly in the building model, but in the mosaik-MATLAB-API. The implemented parallelization allows parametrizing directly in mosaik, whether a connected MATLAB model should be executed in parallel and how much MATLAB workers should be used. The parallelization shows the most impact for simulation models with many instances because the code of the model itself is not parallelized, but only the execution of instances of the model can be distributed on several MATLAB workers.

---

[5] https://mosaik.offis.de

Mosaik provides a scenario API, which allows instantiating the simulation models and a concrete simulation scenario can be defined including data flow between the simulation models. During the execution, mosaik handles the synchronization and data exchange between the models.

9.5 Simulation Scenario and Execution on the Micro-Level

M. Nebel-Wenner, C. Reinhold, J. S. Schwarz

Within NEDS, five transition years (2015, 2020, 2030, 2040, 2050) and three alternatives (decentral, middle ground, central) are considered (as described in detail in Section 8). Each year-alternative combination includes different parameter settings (see, e.g., Table 20 and Table 21).

The maximum grid load (positive and negative) constitutes the most relevant input parameter for the optimized distribution grid planning model (see section 9.7). Hence, the simulations of the smart grid and the smart building model must be structured in a way that this parameter can be determined. However, in order to keep simulation efforts for the coupled smart grid and smart building model within an acceptable range, no simulations of a whole year could be executed. Therefore, we simulated nine different type of days per year-alternative combination to account for seasonal and daily. The different type days result from the combination of the three typical days working day, Saturday, and Sunday with the three seasons summer, winter, and transition. For the estimation of the days with the highest consumption power and the highest generation power, meteorological information was taken into account when determining the type days. It was assumed that the highest power consumption occurs on days with the lowest temperature, especially for the energy consumption of the thermal building supply. On the other hand, the highest production capacity on days with the highest global radiation was assumed for the production of decentralized PV systems. Table 16 accordingly shows the days determined for the respective transition years divided among the type days.

Table 16: Defined data for the type days

Identification code of days	2015	2020	2030	2040	2050
Winter working day (WWT)	10.12.	24.01.	24.01.	24.01.	24.01
Winter Saturday (WSA)	05.12.	05.12.	30.11.	01.12.	10.12.
Winter Sunday (WSO)	04.01.	13.12.	15.12.	16.12.	27.11.
Summer working day (SWT)	02.09.	03.08.	18.06.	03.08.	03.08.
Summer Saturday (SSA)	29.08.	30.05.	03.08.	02.06.	27.08.
Summer Sunday (SSO)	14.06.	06.09.	02.06.	02.09.	18.06
Transition time working day UEWT)	03.04.	24.09.	23.10.	25.09.	01.04.
Transition time Saturday (UESA)	24.10.	28.03.	21.09.	31.03.	22.10.
Transition time Sunday (UESO)	20.09.	19.04.	31.03.	22.04.	25.09.

We considered two example grid settings; one describes a rural setting (land) and the other one a rather urban setting (city). Within the grid sections, agricultural farms, commercial companies, and residential buildings are placed with a distribution in relation to the total building stock of the grid section (shown in Table 17).

Table 17: Distribution of building types in the grid section in %

Building type	Urban setting (city)	Rural setting (land)
commercial companies	10	1
agricultural farms	0	9
residential buildings	90	90

The base load of the commercial companies and the agricultural farms is assigned to a corresponding standard load profile of the groups L and G. Based on distribution assumptions, it is also possible to assign controllability to grid control, PV systems and storage systems to these building types. The electrical behavior of residential buildings, on the other hand, is calculated in detail by modeling the appliances, users and building structures from Section 9.2

The number of buildings in the grid sections is determined based on the maximum load of a local transformer with an apparent power of 630 kVA and a safety range of 10%. This results in a number of 85 buildings for the rural grid area and 64 buildings for the urban grid area for all transition years and alternatives within the scenarios. To reduce the simulation effort, the grid sections were provided with a scaling factor for the number of buildings. This causes only a part of a typical low voltage grid to be implemented. A factor of 2.26 was assumed for the rural grid section and 1.88 for the urban grid section.

Table 18: Percentage distribution of the number of residential units in residential buildings in %

Number of residential units per residential building	Urban setting (city)	Rural setting (land)
1	77.9	67.3
2	19.2	12.1
3	1.16	4.84
4	0.87	3.63
5	0.58	2.42
6	0.29	1.21
7	0	3.36
8	0	2.52
9	0	1.68
10	0	0.84

The load caused on the grid components by the individual buildings is defined by assumed simultaneity factors of the individual appliances from practice. The number of residential units per residential building is as follows.

For each modeled appliance from Section 9.2, the equipment inventory (number of appliances per 100 residential units/buildings) and the static properties are given based on available studies, data sources, and assumptions. As an example, Table 19 shows the equipment inventory for photovoltaic systems in buildings.

Table 19: Equipment inventory for PV-System in number of 100 buildings

Grid section	Alternative	2015	2020	2030	2040	2050
land	middle ground	8.62	13.07	20.49	27.9	35.32
city		2.62	7.07	14.49	21.9	29.32
land	decentral	8.62	17.07	31.16	45.24	59.33
city		2.62	11.07	25.16	39.24	53.33
land	central	8.62	9.055	9.78	10.51	11.23
city		2.62	3.055	3.78	4.51	5.23

Regarding the number of controllable buildings, we assumed that a smart meter including a communication interface is required as described by the attribute *diffusion of ICT standards in the power grid* (see Section 8.2). Since the penetration of such devices varies among the scenarios and simulation years, only a fraction of all buildings at the low voltage grid is part of the VPP. Additionally, it is assumed that not all of those buildings with a smart meter are willing to be part of a VPP. This fraction varies among the three alternatives based on the attributes *share of households participating in demand side management* (see Section 8.2). Table 20 and Table 21 show the setting regarding these two parameters within the different scenarios and alternatives.

Table 20: Share of residential units with a smart meter including a communication interface in %

Scenario	2015	2020	2030	2040	2050
Scenario 1	0,0	5.0	33.9	48.3	61.0
Scenario 2	0.0	2.8	17.5	28.5	32.6
Scenario 3	0.0	2.8	17.5	28.5	32.6
Scenario 4	0.0	5.0	33.9	48.3	61.0
Scenario 5	0.0	5.0	33.9	48.3	61.0

Table 21: Fraction of residential units with a smart meter that is part of the optimization in %

Alternative	2015	2020	2030	2040	2050
decentral	70	70	70	70	70
middle Ground	50	50	50	50	50
central	30	30	30	30	30

Overall, 270 simulations of one day have been executed. From these simulations, the maximum load and the peak feed-in is extracted for each year-alternative-grid combination and passed on to the grid planning model (see Section 9.7). Additionally, the aggregated yearly behavioral adaptation costs for each year-alternative-grid combination are used within the sustainability evaluation.

As described in the previous Section, the models were coupled with the co-simulation framework mosaik, which was also used to define the simulation scenario based on the previously shown parameters. For the execution of the simulation scenario, the infrastructure of the OFFIS institute was used. The simulation was executed in a virtual machine on a server architecture with 12 processor cores and 48 GB RAM. The execution of the 270 simulations took about 2 weeks without parallelization of the building model. The parallelization enables the simulation to use all 12 processor cores and the simulation time decreased to 5 days.

9.6   The Integrated Grid and Market Model

C. Blaufuß

The Integrated Grid and Market Model are developed for a realistic simulation of the continental transmission grid to emulate the European power market and the electric power flow. To enable the goals of the research project NEDS the model is modified to fulfill specific tasks. The following section describes the modifications, the degrees of freedom and the method used for the Integrated Grid and Market Model.

9.6.1   Grid Topology, Power System Delimitation, Interfaces and Degree of Freedom

The transmission and distribution grids represent the electrical power system and form the backbone of the economy and industry as well as the basis for economic wealth in Europe. Figure 56 shows an overview of the electrical power system with the nominal system voltages and the specific system boundaries in Germany.

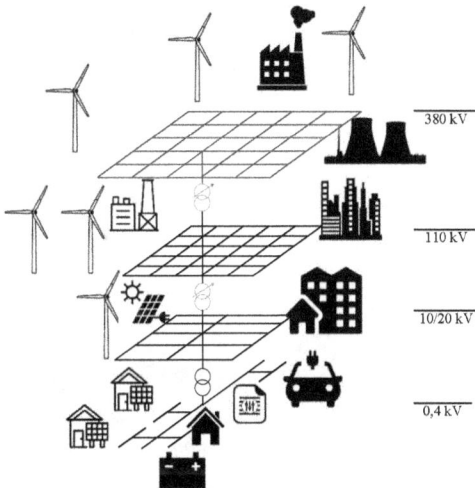

Figure 56: Overview of the electrical power system

The transmission grid is spanned over continental Europe and transports electrical energy over long distances from power plants to areas with high demand. The transmission grid consists of different control zones, which are operated by different independent companies, well known as transmission system operators (TSO). An exchange of power between these zones over interconnectors is possible and depends on feed-in, load condition, and available transmission capacities as well as the power trading price. To minimize the power losses and enable the transport over long distances, the system voltage of the German transmission grid amounts to 380 kV or 220 kV. Large-scale power plants, which are usually thermal power plants, are connected to these voltage levels and provide system services for safe system operation. These services include the provision of active and reactive power for the frequency and the voltage control. The topology of the transmission grid is constructed as a meshed grid. Furthermore, the maximal thermal rating of the transmission elements as well as the voltage bands $U_{min} \leq U \leq U_{max}$ and the short-circuit currents $I''_{k\,min} \leq I''_k \leq I''_{k\,max}$ have to adhere to before and after a failure of one equipment in the grid. A failure of any equipment may not lead to a shutdown of the entire grid. If these conditions are satisfied, the (n-1)-criterion is fulfilled. The nodes of the transmission grid represent the interface to the distribution grids, which consists of the high-, medium and low-voltage grids as well as the respective transformer level. These have different tasks, topologies, and characteristics, which are explained in section 9.7.1.

For the purpose of the project, the model of the existing power transmission system for continental Europe was reduced to the area of the State of Lower Saxony and its neighbors. The transmission lines to grid nodes outside of Lower Saxony are defined as interconnector lines and serves as points for possible power exchange. Figure 57 shows the reduction of the transmission grid and represents the map of Lower Saxony with the transmission lines, power plants and grid nodes used in the reduced transmission system model.

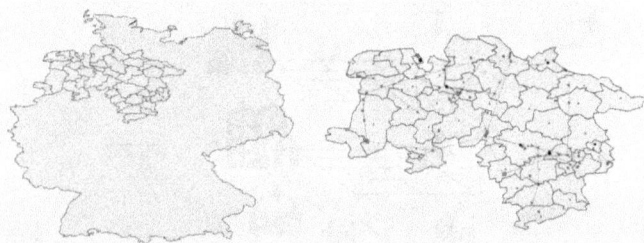

Figure 57: Reduction of the Lower Saxony transmission grid

Red or green lines, representing the 380kV and the 220kV overhead lines, connect the grid nodes. The number of parallel systems is deposited in the database. Furthermore, the colored points show the position, the installed capacity and the type of electrical plants in the map, respectively, in Lower Saxony. Thereby power plants must have a minimum installed capacity to be visible on the map. Power plants can usually be allocated to a specific grid node but in special cases distributed generators, e.g., wind power has to be distributed to different nodes, e.g., the high influence of wind parks in regions. Conventional power plants are able to feed in depending on their marginal costs. Renewable power plants feed in as a priority and their marginal costs are assumed zero. The operation plan of power plants is set for each time step, according to the selectable power plants and their marginal costs. In this connection, the electrical boundary conditions of the grid must be observed, acting as limiting factors for the operating plan. The electrical boundaries consist of the voltage band and the maximal thermal rating. The limit of the voltage maintenance amounts ±10%. Furthermore, the load capacity of the devices depends on their respective type. The visible attributes on the map are a part of the database of the model, which serves as a base for the simulation.

9.6.2 Method of the Integrated Grid and Market Model

The Integrated Grid and Market Model is a compound of two major modules, which are running in a series process. The process starts with the module of the market simulation, receiving the necessary information for the simulation from the database. Another information for the module is provided by the calculation of the energy demand shown in Figure 58. The results of the market simulation contain among other power plant and storage operation plan for every time step in one year. Additionally, to the information contained in the database, this solution also serves as input data for the grid simulation module. The evaluation of a power flow calculation with the operation plan assures that the grid boundaries are adhered to.

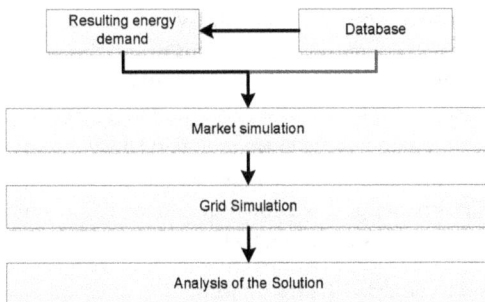

Figure 58: Overview of the integrated Grid and Market Model [166]

In order to generate the necessary criterion for the evaluation of the transition path, an analysis function is required, which is called at the end of the program. A more detail description of the entire algorithm is shown in [166].

9.6.3 Structure and Content of the Used Database

Figure 59 shows an overview of the database of the Integrated Grid and Marked Model, which consists of six major categories. In addition to the above-mentioned information, the power plant data include the rated time series of wind and photovoltaic energy feed-in. The rated time series are required to calculate power the feed-in in every time step. Unlike conventional power plants, renewable plants feed in their power unregulated and do not act on market laws. The category of energy storage almost consists of the same data as the power plant data. Additionally, the attributes of storage capacity and the efficiency factor are added. Since there are only two available forecasted energy storage plants in Lower Saxony, the location and design of future systems represent a degree of freedom. The assumption for the design and the size of storages are based on the studies of the potential of power storage ability in Lower Saxony [5].

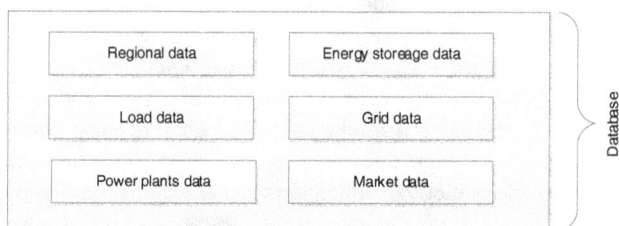

Figure 59: Overview of the database [166]

The load data consist of rated time series for every hour of a year. The loads are scaled with the information from regional data, which include the economic data of the region like population figure, population density and the parts of industry, craft, trade, and households of the consumption. Furthermore, the gross domestic product is used to estimate the electrical consumption, respective, the electric load in the grid for every time step and for every region in Lower Saxony. The load can be allocated to the corresponding nodes. The process of the gross domestic product is emulated by the CEG-model and is shown in Section 9.2.4. Additionally, a more detail description of the database is given by [166].

The grid data is used to describe the transmission grid of Lower Saxony and consists of the selected types of transmission elements as well as the information about the grid topology. Additionally, the interconnectors with limited capacity are included and make a power exchange with neighbored grids possible.

The market data includes economic information like inflation, interest rates, costs of primary fuel and CO_2 certificates as well as the gross domestic product. Especially the data of coast of primary fuel and gross domestic product are difficult to calculate in course of their dependency on world economies. This information is given by the CGE-Model and are updated for every initial year. A more detail disaggregation is described by [166].

9.6.4 Market Simulation

The market model emulates the electricity market and is based on the calculation of the marginal costs of the conventional power plants and energy storages in the defined area, to cover the energy demand. Part of the important input data for the market model is represented by the load time series. This data is calculated by the sum of feed-in power of the renewable energy sources and the consumption of the loads for every time step. Furthermore, the value could be positive or negative.

$$P_{Residual} = P_{Load} + P_{Renew} \qquad (9\text{-}5)$$

The value of the residual power $P_{Residual}$, resulting from the assumption of priority feed-in of the renewable power plants has to be equalized by the conventional power plants, the power storages, and the imported power as well as the exported power.

$$P_{Residual} + P_{Stor} + P_{Plants} + P_{Exp} = 0 \qquad (9\text{-}6)$$

On the one hand, it is possible to balance a positive value of $P_{Residual}$ by every element of the equation (9-6). On the other hand, a negative value of $P_{Resiudal}$ can only be covered by the storage systems or a power export.

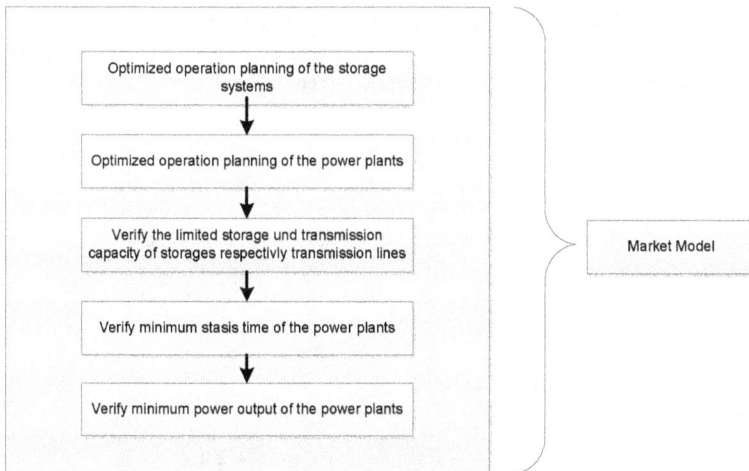

Figure 60: Overview of the market model [166]

Thereby, the constraints of the energy capacity of storages and transmission capacity of the interconnector as well as the minimum output and stasis time of the power plants, have to be observed. Figure 60 shows a more detailed sequence of the market model. Both operation planning functions calculate their solution by solving an integer linear optimization problem (ILP) with a commercial solver while obtaining the boundary conditions, as shown in Figure 60.

A more detailed description of the applied optimization functions is shown in [166], [167]. The output of the described module delivers an optimized operation plan of the connected power plants and energy storage systems. An example of an optimized operation plan is shown in Figure 61.

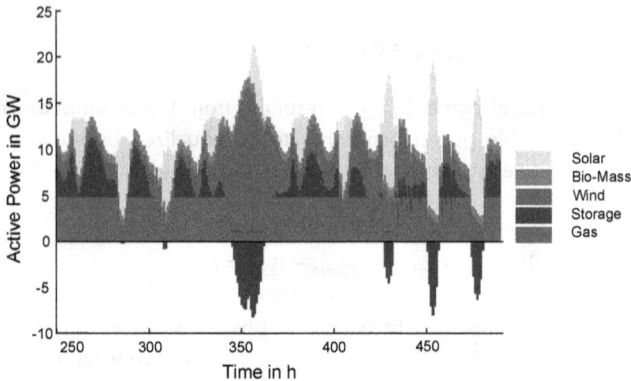

Figure 61: Example of an optimized operation plan

Additionally, the model generates market-based solutions to provide other particular projects with information or to evaluate the system condition.

9.6.5 Grid Simulation

The following grid simulation model mainly consists of a power flow calculation, based on the load assumption and the operation plan. The calculations are applied for every time step of the considered year. The node voltages, the load of the transmission elements are used to detect the requirements of grid reinforcements. A detailed explanation is shown in [166].

9.6.6 Application of the Integrated Grid and Market Model to NEDS

The aim of the calculation is to represent different alternatives of the future transmission grid in Lower Saxony and their transition path with respect to coping with the future load and feed-in assumptions as well as a changing power plant and storage system park. The different alternatives of the scenario can be read in Section 8. The results of the target states and transition paths are used to grade the alternatives of the scenario in Section 11. The starting point for every simulation is given by the configuration of the Lower Saxony transmission grid in the year 2015 (Figure 62) and is additionally used to compare the future development of the grid.

The Integrated Grid and Market Model are used to calculate the target states and the transition paths into different ways.

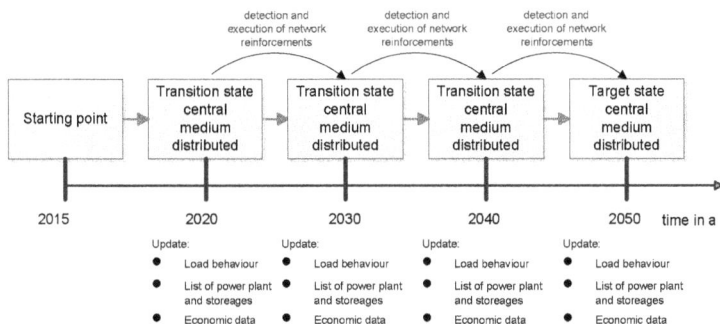

Figure 62: Approach to determine the target state and transition path

In the first step, the target states are modeled and calculated for three different alternatives to the chosen scenario. The required data for the calculations are obtained by the developed scenarios, respectively, the alternatives of them, as well as the information, exchanged with the CGE-model.

In the second step, the calculation of the transition paths occurred for the year 2020 2030 and 2040. Therefore, a separate transition path is calculated for every alternative of target states. Before the simulation executes for the specific year, the required data is updated (Figure 62). The power flow calculation contained in the Integrated Grid and Market Model checks congestion problems in the grid. Additionally, possible congestions are compensated by reinforcement measures and are executed before the simulation of the next time step starts.

The calculated criteria, e.g. energy mix, grid efficient and CO_2 –emission are used to evaluate the transition path or serve as input data for the Computable General Equilibrium Model (section 9.8). Furthermore, the results of the calculation are shown in Section 10.2.

9.7 Optimized Distribution Grid Planning

C. Blaufuß

In NEDS, a grid-planning algorithm developed by the Institute of Electrical Power Systems of the Leibniz Universität Hannover emulates the distribution grid. The program plans optimized grids, observing future load and supply behavior as well as a boundary condition.

9.7.1 Grid Topology, Power System Delimitation, Interfaces and Degree of Freedom

The distribution grid consists of the high-voltage grid, the medium-voltage grid and the low-voltage grid as well as the respective transformer levels. An overview of the distribution grid and it is connection with the transmission grid is shown in Figure 56.

The high-voltage grid transports energy in local consumption areas and function as a connecting point for large loads (steel plants) and power plants (wind parks) as well as an interface to the medium voltage grid. The system voltage mainly in Germany amounts to 110kV. The high-voltage-level also has to fulfill the (n-1) criterion resulting in a soft intermeshed grid topology.

Medium-voltage-grids are connected to the high-voltage-grid by transformers. They supply low-voltage-grids, loads with high consumption, single wind turbines and large photovoltaic power plants. The voltages range from eg. 10 kV, 20 kV to 30 kV in Germany. Generally, the normal system voltage of rural MV grids amounts to 20 kV. Furthermore, urban grids operate with a normal voltage of 10 kV. A 30 kV voltage level represents a special case, serving to connect industrial parks. These grids use the higher voltage level to limit the current, caused by the requirement of a high amount of power.

In contrast to the higher voltage levels, the medium-voltage grid only has to hold the attenuated form of the (n-1)-criterion. A failure of a transmission element can lead to a shut down in a line of a medium-voltage-grid. However, after clarifying the fault, the operator has to repower the affected line by switch operation. Therefore, the medium-voltage grids are designed as a ring or soft meshed grids, operating with open separation points. Furthermore, the (n-1)-criterion has only hold on be fulfilled for loads. Power plants have no claim to feed in case of a failure.

The low-voltage (LV) grid is connected to the medium-voltage grid by local substations. It supplies households and small industries. A change appears in this area, due to a new load behavior as well as feed-in behavior. New loads are integrated, e.g., heat pumps and charging stations for electrical vehicles. Additionally, the number of photovoltaic systems mounted on rooftops move on. The low-voltage-grid has a radial layout and does not fulfil the (n-1)-criterion. An exception is represented by urban LV grids, which are constructed as a soft meshed grid.

The grid-planning algorithm generates optimized grid designs for the high- and medium-voltage-level. Furthermore, both grid levels are optimized contiguously in an iterative algorithm and include the HV/MV transformer levels. The transmission grid and the lower-voltage grid represent the system delimitations for the planning algorithm, which is shown in Figure 63.

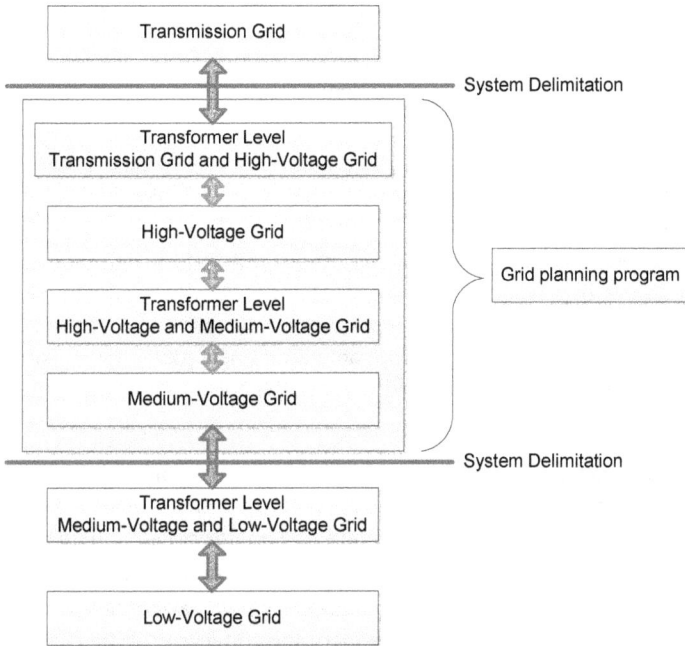

Figure 63: System delimitations for the planning algorithm

The interface to the transmission grid is determined by its node voltage. Additionally, the interfaces to the low-voltage grids are emulated by load and feed-in behavior. The behavior is provided by subprojects 3, 4 and 5 and is combined to layout scenarios. The fixed position of the grid nodes is an assumption of the developed program, which is justified by a constant location of cities, villages and companies. However, a change in the number of grid nodes over the years is possible, considering the building of new wind parks or districts. Table 22 shows the boundary conditions for the planning and optimization of high- and the medium-voltage grids. Different boundary conditions have to be complied with, depending on the voltage level.

Table 22: Boundary conditions for the high- and the medium-voltage grid

category	High-voltage grid	Medium-voltage grid
(n-1)-criterion	Yes	Yes (switch)
Voltage bands	±10 %	±10 % or ±5 %
Short-circuit current		
Min. short-circuit current	depends on primary equipment (max. Impedance)	depends on primary equipment (max. Impedance)
Max. short-circuit current	depends on primary equipment (min. Impedance)	depends on primary equipment (min. Impedance)
Max thermal rating	depends on the relative transmission element	depends on the relative transmission element

The voltage drop in the high-voltage level is not firmly defined and assuming various voltage values depending on the specifications of the consumer or subordinated grids. Therefore, the assumption of boundaries of 0.9 per unit (p.u.) and 1.1 p.u. is implemented that the voltage drop on every single point must not be over, respectively, under 10 % of the rated voltage. This specification represents a degree of freedom in grid planning.

The medium-voltage and low-voltage levels have a common MV/LV-range, defined from 0.9 p.u. to 1.1 p.u. without using a voltage control of the transformers. The voltage boundaries for the medium-voltage grid approximately add up to ±5 % as a result. Furthermore, the voltage drop amounts up to ±2 % for the MV/LV transformer level. Finally, 3 % of the voltage drop is earmarked the low-voltage grid. When using voltage controlled transformers, the grid levels can be decoupled. As a result, both grid levels are able to use the full voltage range of ±10 % [168], [169].

The categories of short circuit current and transmission element overload depend on the specific grid protection devices and the overload behavior of the used types of transmission elements. Therefore, no general statement is possible and must be checked for every single event [170].

9.7.2 Method of the Grid-Planning Algorithm

The grid-planning algorithm acts in an iterative process and consists of two main modules with different optimization goals. The operation chart of the planning program is shown in Figure 64. The first module is realized by a heuristic algorithm, optimizing the topology of the grid and represents the main construct of the program. The second module, named Grid Reinforcement Module, is integrated into the heuristic algorithm. The position of the module is shown in Figure 64 (red). It configures the transmission elements of the medium-voltage grid by solving an integer linear optimization problem (ILP).

Furthermore, the high- and the medium-voltage grid, as well as the transformer level, are planned overarchingly. The combined planning of the levels enables more degrees of freedom and extends the feasible space. The consideration of existing grid structures represents another feature of the planning program and is essential to calculate the transition paths of the considered grid.

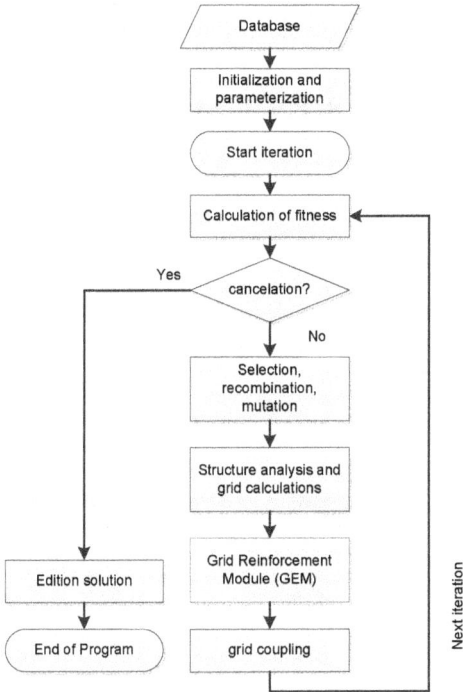

Figure 64: Operation chart of the planning program

9.7.3 Structure and Content of the Used Database

The required database consists of electrical data as well as geographic and economic data. The composition of the database is shown in Figure 65.

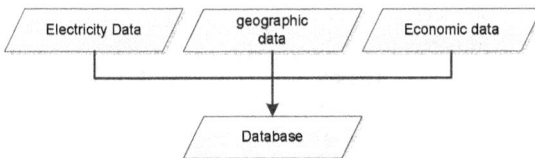

Figure 65: Composition of the database

The electric data includes the maximum load and feed-in assumptions for different states as well as the equipment data of the possible selectable transmission elements. Furthermore, the electric data contains information about the characteristic of the respective equipment. The geographical data includes the position of the substations within the regarding the area and their distances to each other. Any kind of costs is located in the economic data, representing, e.g., the costs of equipment, power losses and the cost of properties for substations.

9.7.4 Mathematical Description of the Grid Topology

The connection within the grids and their transformer levels are described mathematically using matrices. Therefore, an example of a grid and a transformer topology is shown in Figure 66. The matrix A_T represents a general description of the transformer level. The first column A_T describes the station number, which connects the grid to the higher voltage level using a transformer. The digit in the second column contains information about the used transformer type. Furthermore, the third element shows the number of parallel transformers. The last column displays the position of the transformers tab.

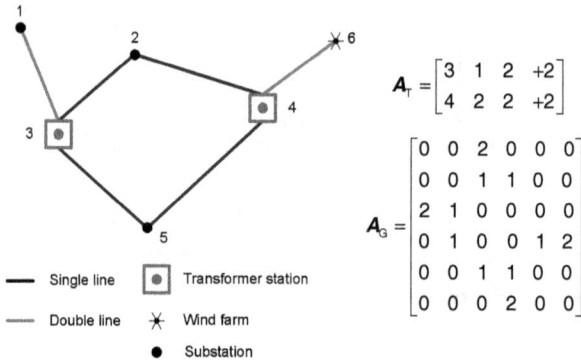

$$A_T = \begin{bmatrix} 3 & 1 & 2 & +2 \\ 4 & 2 & 2 & +2 \end{bmatrix}$$

$$A_G = \begin{bmatrix} 0 & 0 & 2 & 0 & 0 & 0 \\ 0 & 0 & 1 & 1 & 0 & 0 \\ 2 & 1 & 0 & 0 & 0 & 0 \\ 0 & 1 & 0 & 0 & 1 & 2 \\ 0 & 0 & 1 & 1 & 0 & 0 \\ 0 & 0 & 0 & 2 & 0 & 0 \end{bmatrix}$$

— Single line   ⊡ Transformer station

—— Double line   ✳ Wind farm

● Substation

Figure 66: Mathematical description of a grid and a transformer topology

The matrix A_G, well known as the adjacency matrix, contains information about the particular connections between the single stations in the grid. The matrix has a n × n dimension with a symmetric structure. Furthermore, the numbers of rows, respective, columns of the matrix represent the number of stations in the grid. The matrix is weighted with positive integer values. Additionally, the value represents, which transmission elements are used. Zero stands for no connection. Furthermore, a value unlike zero is dedicated to different types of transmission elements. In conclusion, the matrixes A_G and A_T are used to describe every individual grid design.

9.7.5 Topology Optimization Module

The topology optimization module is based on a heuristic and combines a genetic algorithm with a local search and a simulated annealing algorithm. The main heuristic is represented by a genetic algorithm and consists of the functions selection, recombination, mutation and fitness calculation. The algorithm uses evolutionary strategies to find an optimized topology to adapt with load and feed-in assumptions. This improvement is accomplished by applying laws of nature. Therefore, individual grid designs are combined in a pool, well known as population, consisting of a defined number of grid designs.

The individuals of the population are evaluated by the value of the objective function and favored to survive for small corresponding values. The decision about the survival of the particular individuals of populations is given in the function mating pool. The populations are divided into groups consisting of four individuals. In result, the best of the four individuals survive and take part in the new mating pool. Subsequently, the individuals of the mating pool represent the parents, creating the children of the new population, which are the input parameter for a possible next iteration step. This progress occurs in the function recombination and consists of a combination of cross over and local search features. Figure 67 shows an example of a crossover feature and Figure 68 represents an instance of the local search function.

The crossover feature randomly selects two individuals of the population, called parents and divides them at the same column into two parts. Furthermore, the first part of parent one and the second part of parent two are paired with each other, forming the child one of the next population. The other remaining parts of the parents form child two. The algorithm uses several crossover features differs in the number of cuts. The selection of cutting positions occurs always randomly. The local search feature uses information from the structure analysis function to improve grid design. Two examples are shown in Figure 68. The first example is given by parent 1. The stations two and five of parent 1 are doubly connected to each other. The first connection occurs directly and the second connection is indirectly realized over stations three and four. The used feature of the local search deletes the direct connection and realizes only the indirect connection. Additionally, the second example is explained with parent 2. In case of a crossing of two transmission lines, the function changes the topology, causing the crossing to disappear. The result of the topology change is represented in the modified topology of child 2. Both variants of the local search are aimed at improving the topology. Furthermore, an improvement in the topology does not always lead to the improvement of the entire grid design due to the possible change in the transmission elements types [171].

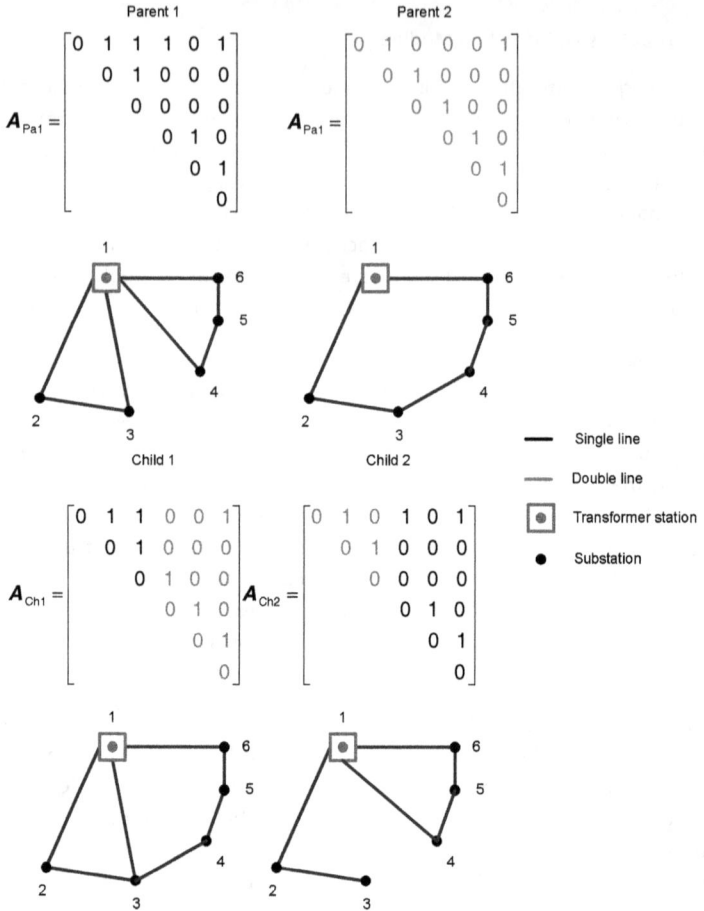

Figure 67: Example of cross-over function

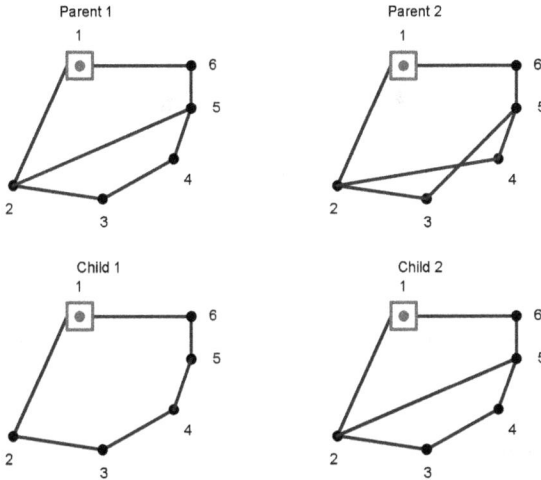

Figure 68: Example of the topology change

In the further course, the mutation function is applied, which manipulates the grid designs and creates a certain rate on deviation in the population.

The child 2 in Figure 67 shows that the recombination may lead to impermissible grid topologies and is visible in the single line connection of station two and three. The recombination can be led to the disintegration of the grid into several parts. An analysis of the topology checks the grid design against such failures. A special and difficult problem is represented by the connection of grid parts with a single transmission line. A failure of the single line may lead to a shutdown of a part of the grid that cannot be repowered by switching operations. This problem is well known as Konigsberg Bridge Problem and cannot be solved analytically. The problem is solved by an algorithm, which reduced the adjacency matrix in an iterative process. The reduction combines stations with only two connections into stations with more than two connections, shown in Figure 69.

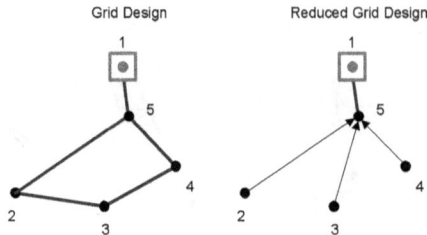

Figure 69: Reduction of the adjacency matrix

In result, bridges are shown in the reduced adjacency matrix as a line with only one connection. Finally, the necessary separation points and the input data for the grid reinforcement module are determined by power flow calculations.

The functions of the mating pool, recombination, and mutation, as well as fitness calculations, build the heuristic algorithm, optimizing the topology of the grid designs. Additionally, the simulated annealing is applied, preventing a premature convergence in local optima by tolerating the worsening of the solution [171].

9.7.6 The Grid Reinforcement Module

The choice of transmission element types occurs in a different way depending on the voltage level. The choice of the transmission element types of the high voltage level is optimized by switching and changing the numbers in the adjacency matrix by the heuristic algorithm, due to the nonlinear relationship between the impedance and power flows. The impact and the evaluation of the change of transmission element types are observed by grid calculations and serve at ones to hold on the grid boundaries.

The optimized planning process of the transmission elements in the medium voltage grid is different from the planning process in the high voltage grid. The topology of the medium-voltage grid is planned as a ring or intermeshed grid and operates with opened separation points. Additionally, the medium-voltage also operates as a radial grid in case of a disturbed condition. The affected transmission element is switched off and the line is repowered by closing the respective separation point. The length of the line increases and represents one of the possible worst grid conditions.

The grid reinforcement module optimizes the choice of the transmission element types for these conditions and applies two important load and feed-in scenarios to reinforce the grid. The first scenario represents the maximum amount of load while decentralized plants are inactive. The second scenario switches the behavior of load and feed-in. It consists of maximum feed-in of decentralized plants and the minimal load's acceptance. Figure 70 shows the applications of the scenarios to the respective grid conditions.

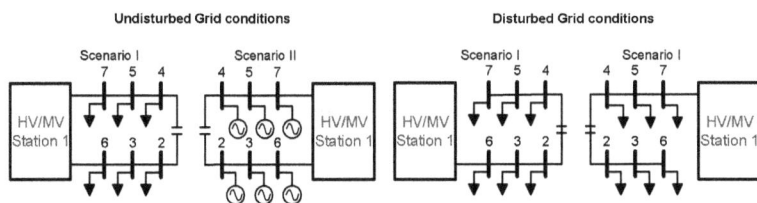

Figure 70: Used grid scenarios and grid conditions for the layout of the transmission element

The first grid scenario is applied to the undisturbed and disturbed grid conditions, observing the security of the power supply. The second scenario is only applied to undisturbed grid conditions. In case of a disturbed grid condition, decentralized power plans are fully limited in their power supply and do not feed-in their power to the grid. Under consideration of the specified scenarios and grid conditions, all electric boundaries must be satisfied.

The possible choice of transmission element types can be described as a linear optimization problem calculated by a commercial solver due to the radial layout of the MV grid in case of the undisturbed and disturbed grid condition. Figure 71 shows the operation chart of the Grid Reinforcement Module and it is embedding in the grid planning algorithm.

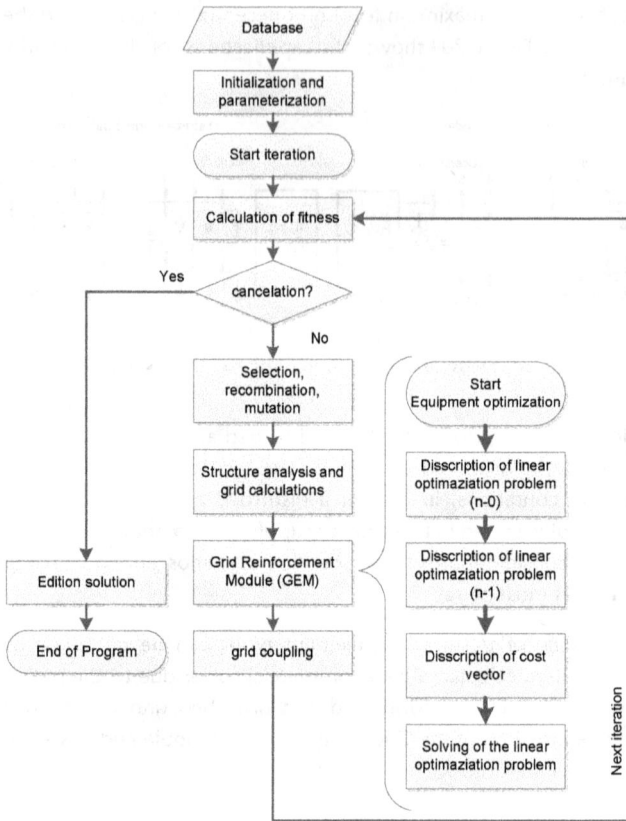

Figure 71: Overview of the grid reinforcement module

The approach to use an integer linear optimization problem (ILP) for the layout of transmission element types in a grid extension algorithm for the low voltage grid is described in [172]. The ILP functions as the foundation of the grid reinforcement module (GEM). The GEM provides an optimal choice of transmission element types regarding the undisturbed and disturbed grid condition. Additionally, load and feed-in scenarios are considered. The aim of the GEM is to find a solution in compliance with the electrical criteria (section 9.7.1) while minimizing the costs. The application of the grid reinforcement module is described by an example shown in Figure 72.

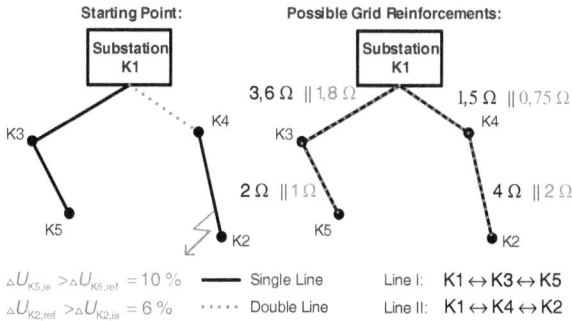

Figure 72: Example for the grid reinforcement module

The example consists of four secondary substations connected to the HV/MV substation. The voltage drop on station K3 exceeds the given criterion of 10 % and makes grid reinforcements necessary. Only the undisturbed operation is considered, using the scenario I, to keep the example simple.

The ILP represents a subtopic of mathematics and deals with the optimization of linear objective functions. The feasible set is limited by linear equations and inequality conditions. Furthermore, the solution vector must partially provide a binary vector and represents a special field, the Integer Linear Optimization (ILP). The formulation of the optimization problem is represented by equation (9-7).

$$\min_{x} f^{T} \cdot x \text{ subject to} \begin{cases} x \text{ (intcon) are integers} \\ A_{ineq} \cdot x \le b_{ineq} \\ A_{eq} \cdot x = b_{eq} \\ lb \le x \le ub \end{cases} \tag{9-7}$$

The length of the solution vector x is given by the numbers of parallel transmission elements (single lines or double lines) and the number of connection in the grid. Therefore, the solution vector provides a binary solution ($x_i = 1$ equipment i is installed; $x_i = 0$ equipment i is not installed). Additionally, the costs of the transmission elements for each connection are represented within f. Both vectors, x_i and f form the objective function and is shown by equation (9-8).

$$\min f^{T} \cdot x \tag{9-8}$$

An example of the objective function is given by equation (9-9), corresponding to Figure 72.

$$\min f^T \cdot x =$$

$$[\alpha_{K1\leftrightarrow K3} \quad \alpha_{K1\leftrightarrow K4} \quad \alpha_{K2\leftrightarrow K4} \quad \alpha_{K3\leftrightarrow K5}] \begin{bmatrix} \beta_{K1\leftrightarrow K3} \\ \beta_{K1\leftrightarrow K4} \\ \beta_{K2\leftrightarrow K4} \\ \beta_{K3\leftrightarrow K5} \end{bmatrix}$$

(9-9)

with:

$$\alpha_{Kx\leftrightarrow Ky} = \begin{bmatrix} K_{\epsilon, Kx\leftrightarrow Kx, 1} & \cdots & K_{\epsilon, Kx\leftrightarrow Kx, n-1} & K_{\epsilon, Kx\leftrightarrow Kx, n} \end{bmatrix}$$

$$\beta_{Kx\leftrightarrow Ky} = \begin{bmatrix} x_{Kx\leftrightarrow Kx, 1} & \cdots & x_{Kx\leftrightarrow Kx, n-1} & x_{Kx\leftrightarrow Kx, n} \end{bmatrix}$$

n: number of parallel transmission elements

The solution of the optimization problem represents the configuration of transmission elements with minimal costs while maintaining the grid restrictions. A solution of the ILP, hence a design of the transmission elements is including the entire grid design. Therefore, the solution vector **x** must provide a result for each planed transmission line, which is ensured by the equation of the ILP. The general form of the equation is represented by equation (9-10).

number of planed tranmission elements

$$A_{eq} \cdot x = b_{eq} = \begin{pmatrix} d & & & 0 \\ & d & & \\ & & \ddots & \\ 0 & & & d \end{pmatrix} \cdot x$$

(9-10)

number of transmission types

with $d = \begin{bmatrix} 1 & \cdots & n\text{-}1 & n \end{bmatrix}$

n: number of parallel transmission elements

The equation of the ILP is shown by equation (9-11), regarding to the example of Figure 72.

$$A_{eq} = \begin{array}{c} \begin{array}{cccccccc} K1 \leftrightarrow K3 & K1 \leftrightarrow K4 & K2 \leftrightarrow K4 & K3 \leftrightarrow K5 \end{array} \\ \begin{bmatrix} 1 & 1 & 0 & 0 & 0 & 0 & 0 & 0 \\ 0 & 0 & 1 & 1 & 0 & 0 & 0 & 0 \\ 0 & 0 & 0 & 0 & 1 & 1 & 0 & 0 \\ 0 & 0 & 0 & 0 & 0 & 0 & 1 & 1 \end{bmatrix} \end{array} ; b_{eq} = \begin{bmatrix} 1 \\ 1 \\ 1 \\ 1 \end{bmatrix}$$

(9-11)

The vector b_{eq} of the equation is set to ones to guarantee that each planned connection is realized with one of the possible transmission elements.

Generally, the voltage range and minimal short-circuit current criteria are the most relevant restrictions, due to high distances in case of fault occurrence. However, the transmission elements are checked for overloads by a single row of the inequality condition of the ILP. All values are initialized with zero at the beginning of the module. In the case of undesirable current values above the allowed thermal limit of the equipment, the respective element is set to a positive value. The vector $\mathbf{b}_{\text{ineq}}$ is set to zero preventing the transmission element to be chosen. As a result, only transmission elements observing the thermal limit are selectable. In conclusion, the inequality condition ensures compliance with the voltage boundaries and the minimum short-circuit currents. The inequality condition is shown by the following equation.

$$\mathbf{A}_{\text{ineq}} \cdot \mathbf{x} \leq \mathbf{b}_{\text{ineq}} \tag{9-12}$$

The matrices $\mathbf{A}_{\text{ineq}}$ and $\mathbf{A}_{\text{eq}}$ have an equal number of columns and are defined analogously to (9-12) differing in the number of rows. The number of rows of the matrix $\mathbf{A}_{\text{ineq}}$ is equal to the sum of the number of a single line of the radial layout grid and the number of their applied scenarios. The elements of the matrix $\mathbf{A}_{\text{ineq}}$ represent the impedance values of individual transmission elements of the realized connections. The general form of the matrix according to the example of Figure 72 is shown by equation (9-13).

$$\mathbf{A}_{\text{ineq}} =$$

$$\begin{bmatrix} \delta_{K1\leftrightarrow K3} & \delta_{K1\leftrightarrow K4} & \delta_{K2\leftrightarrow K4} & \delta_{K3\leftrightarrow K5} \\ \chi_{K1\leftrightarrow K3} & 0 & 0 & \chi_{K3\leftrightarrow K5} \\ \chi_{K1\leftrightarrow K3} & 0 & 0 & \chi_{K3\leftrightarrow K5} \\ 0 & \chi_{K1\leftrightarrow K4} & \chi_{K2\leftrightarrow K4} & 0 \\ 0 & \chi_{K1\leftrightarrow K4} & \chi_{K2\leftrightarrow K4} & 0 \end{bmatrix} \begin{matrix} \text{overload} \\ \text{String I } \Delta U_{K5} \\ \text{String I } I''_{k,\min,K5} \\ \text{String II } \Delta U_{K2} \\ \text{String II } I''_{k,\min,K2} \end{matrix}$$

with:

$$\chi_{Kx\leftrightarrow Ky} = \begin{bmatrix} Z_{Kx\leftrightarrow Ky,1} & \cdots & Z_{Kx\leftrightarrow Ky,n-1} & Z_{Kx\leftrightarrow Ky,n} \end{bmatrix}$$

$$\delta_{Kx\leftrightarrow Ky} = \begin{bmatrix} \vartheta_{Kx\leftrightarrow Ky,1} & \cdots & \vartheta_{Kx\leftrightarrow Ky,n-1} & \vartheta_{Kx\leftrightarrow Ky,n} \end{bmatrix}$$

$$\vartheta = 0 \,\|\, 5$$

n: number of parallel transmission elements

$$\tag{9-13}$$

The product of the matrix $\mathbf{A}_{\text{ineq}}$ and the solution vector $\mathbf{x}$ represent the total impedance of the respective line. Furthermore, the lines must be observed a particular impedance value, depending on the voltage range and the minimal short-circuit current. The maximum impedance value is determined by the method of angular momentum, to maintain the voltage restrictions. The node voltages result from a power flow calculation. The maximum impedance value depending on the node voltages is calculated in equation (9-14).

$$Z^{\Delta U}_{Kx \leftrightarrow Ky, max} = \frac{\Delta U_{Ky,ref}}{\Delta U_{Ky,is}} \cdot (Z_{Kx \leftrightarrow Ky,is}) \tag{9-14}$$

The maximum impedance value depending on the short current criterion is shown by equation (9-15).

$$Z^{k,min}_{Kx \leftrightarrow Ky, max} \leq Z_{line, max} - (Z_{1Tran} + Z_{2Tran}) - (Z_{1Grid} + Z_{2Grid}) \tag{9-15}$$

The maximal impedance value is calculated for the lines with and without violations of the voltage criterion. The possibility of improvement exists in healthy lines. The limiting factor of the impedance is represented either by the voltage range or by the short-circuit limit. Hence, the ILP is also enabled to selects transmission elements that increase the line impedance. Both criteria must be implemented. The entire inequality condition is shown in equation (9-16).

$$\begin{bmatrix} \delta_{K1 \leftrightarrow K3} & \delta_{K1 \leftrightarrow K4} & \delta_{K2 \leftrightarrow K4} & \delta_{K5 \leftrightarrow K5} \\ \chi_{K1 \leftrightarrow K3} & 0 & 0 & \chi_{K3 \leftrightarrow K5} \\ \chi_{K1 \leftrightarrow K3} & 0 & 0 & \chi_{K3 \leftrightarrow K5} \\ 0 & \chi_{K1 \leftrightarrow K4} & \chi_{K2 \leftrightarrow K4} & 0 \\ 0 & \chi_{K1 \leftrightarrow K4} & \chi_{K2 \leftrightarrow K4} & 0 \end{bmatrix} \cdot x \leq \begin{bmatrix} 0 \\ Z^{\Delta U}_{K1 \leftrightarrow K5, max} \\ Z^{k,min}_{K1 \leftrightarrow K5, max} \\ Z^{\Delta U}_{K1 \leftrightarrow K2, max} \\ Z^{k,min}_{K1 \leftrightarrow K2, max} \end{bmatrix} \tag{9-16}$$

The solver takes part in the grid reinforcement module. The commercial solver Gurobi is used.

9.7.7 Application of the Grid-Planning Algorithm to NEDS

The goal of the calculation constitutes the planning of future grids and their transition path with respect to coping with the future load and feed-in assumptions. As a result, the grid-planning program calculates optimized grid designs. The sequence of calculation is shown in Figure 73.

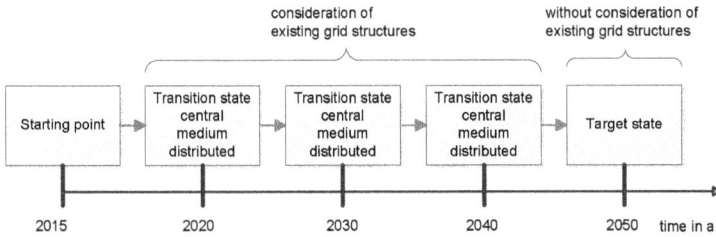

Figure 73: Approach to determind the target state and transition path

The grid of the target state is planned without the consideration of existing grid structures and grid elements, justified by the long term and thereby the depreciation of all equipment. The specific transition path takes into account the existing infrastructure of the grid. An update of the load and feed-in behavior occurs between the relative time steps receiving from the other subprojects. Furthermore, the power infeed's of wind farms are integrated.

9.8 Computable General Equilibrium Model

H. Krause, M. Hübler

As a contribution to the literature, the new forward calibration process described by Pothen & Hübler [173] enables the adjustment of the region- and sector-specific productivities within a new quantitative trade theory model, particularly a Ricardian Eaton and Kortum model [174]. This allows the modeler to replace the standard Armington approach to modeling international trade in a recursive-dynamic long-term simulation framework [175] by an advanced approach with micro-foundations [176]. The main advantage of the forward calibration process is that researchers can impose various constraints on economic growth and on structural change at the same time [173]. This saves time and is more accurate and transparent as compared to standard approaches, where combinations of constraints have to be imposed manually; further, the model can be calibrated to a variety of different scenarios, once the forward calibration process is implemented [173].

One of the key factors of the future scenarios is climate and energy policy plans describing national and international developments (see Section 6). Climate and energy policies are established at the regional, national and international level. In a globalizing world with international trade in the energy sector, energy policies and trade policies at different levels influence each other and thus should be analyzed jointly. Most climate and energy policies target climate protection through CO_2 reductions, particularly the European Union's emissions trading system (EU ETS).

For the analysis of energy policies affecting Lower Saxony's (LSX) (or alternatively Northwest Germany's (NWG)) economy, in the project, we have constructed a static Ricardian general equilibrium model described by Pothen & Hübler [176] following [174, 177, 178, 179] with one representative consumer per region and representative firms in all regions and sectors. The development of LSX's or NWG's economies is compared to the rest of Germany, because they are closely linked via national energy policies. We combine the EK approach with nested constant elasticity of substitution (CES) functions and a disaggregated electricity sector to analyze climate policies with the goal of substantially increasing the renewable energy shares in the electricity sector [180, 181, 182, 183]. We further assume constant returns to scale, perfect competition (zero profits) and clearance of all markets.

Our model allows the analysis of Northern German, German and European climate and energy policy in a globalizing world with intense international connections via international trade. The analysis is divided into two steps: in a first step, we study the influence of climate policy, trade policy and their interconnections on the energy sector and the economy of LSX for the benchmark year 2011. In a second step, we translate the future scenarios (FS), as described in Section 6.2, to our economic framework and implement them in the model to capture macroeconomic developments and political conditions of NWG and the other model regions from 2011 until 2050.

In the first step, we simulate two global trade liberalization scenarios and two climate policy scenarios and additionally combine the two trade policy scenarios with the first climate policy scenario [176]. (These scenarios describe the status quo at the baseline (2011) and are not related to the scenarios described in Section 6.2.) For this analysis, we define the two regions Lower Saxony (LSX) and the rest of Germany (ROG₁), which comprises all other states in Germany except LSX. Energy policies at the European Union (EU), Germany and Lower Saxony level are taken into consideration.

The model uses the same input-output data for the structural estimation procedure of the Eaton and Kortum (EK) module as for the general equilibrium model calibration. The EK trade module is calibrated by extending the structural estimation procedure [184, 185] by introducing a market clearing condition and iceberg trade costs in the gravity model. Further, our model overcomes the assumption that the service sector and the electricity sector produce non-tradeable goods, and we can calculate explicit estimates of productivities and iceberg trade costs. Those estimates are needed for the trade policy simulations and are an improvement as to the current state of CGE models in this field [177]. An elaborated sensitivity analysis evaluates the robustness of the results. See Pothen & Hübler [176] for a detailed discussion of the model and the appendix of the article for a robustness check and a sensitivity analysis.

In the trade policy scenarios, we investigate the effect of global trade costs on welfare, CO_2 emissions and the electricity mix in LSX and ROG₁. Two trade policy scenarios are created [176]: The *noTarrifs* scenario and the *lessIceberg* scenario. In the *noTarrifs* scenario, we abolish all import tariffs and export subsidies in all regions and sectors, to imitate the effect of a multilateral trade agreement under the auspices of the World Trade Organization (WTO). In the *lessIceberg* scenario, we reduce non-tariff trade barriers that occur because of different regulations on products in all countries. To calculate productivities and iceberg trade costs on a disaggregated regional and sectoral level, we introduce a log-multiplicative specification of the gravity model [186, 187].

The major difference between tariffs and iceberg costs is that iceberg costs are mere inefficiencies in trade, whereas tariffs generate revenues that are transferred to the respective consumer [176]. Thus, the reduction of tariffs may lead to smaller welfare gains than the reduction of iceberg costs. To make the two scenarios comparable, iceberg costs are reduced to the extent that the reduction on overall trade costs is the same as in the *noTarrif* scenario.

For the climate policy scenarios, we assess (i) the effect of a tighter CO_2 emissions cap and renewable energy support on welfare and (ii) CO_2 emissions and the electricity mix in LSX and ROG_1 [176]. The two policies are in line with the official targets of the European Commission. The CO_2 emission cap is currently implemented with the European Union's emissions trading system (EU ETS), whose target is a 13% reduction of CO_2 emissions vis-à-vis the emissions in the year 2011. This corresponds to a 21% reduction vis-à-vis 2005 via the EU ETS as envisaged by the European Commission [188] for the year 2020.

The EU ETS basically generates a market for allowances to emit CO_2. Those allowances can be traded by actors in the sectors covered by the scheme, e.g., the power sector. Sectors that are not covered by the EU ETS and end-customers are not affected by the scheme. In the first climate policy scenario (scenario ETS), climate policy is implemented as the EU ETS only. The second climate policy scenario implements the EU ETS and adds support for renewable energies (RES) for wind power, solar power, geothermal energy, and biomass. The National Renewable Energy Action Plans (NREAP) [189] guides in the targets for shares of renewable energy in the power generation mix.

Table 23: Parameters used to define future scenarios (FS)

Parameter	Scenario dependent	Implementation form
The growth rate of GDP	Yes	Directly
Change in the share of agriculture in value added	Yes	By restrictions
Change in the share of services in value added	Yes	By restrictions
Capital stocks	Yes	Directly
Prices of primary energy carriers	Yes	Directly
Autonomous energy efficiency improvements	Yes	Directly
Share of technology g in electricity generation (NWG only)	Yes	Directly
European emissions reduction goals	No	By restrictions
Share of freely allocated allowances in the EU ETS	No	By restrictions
Implicit CO_2 price in sectors outside the EU ETS	No	Directly
Labor endowments	No	Directly
Productivity change in generation technologies	No	Directly

In the second step of our analysis, we define future scenarios (FS) as described in Section 6.2 and implement them in the model [173]. The focus region in this step is expanded to Northwest Germany (NWG) including Lower Saxony, Hamburg, and Bremen, based on the results of the discussion at "Roundtable Energy Transition" (Runder Tisch Energiewende). The rest of Germany (ROG_2) comprises all other states of Germany. Overall, there are five future scenarios (FS) (see Section 6.2) based on twelve parameters. These parameters are either implemented directly or enforced by restrictions (s. Table 23). The investigated time frame reaches from 2011 until 2050. Details on the data used to calculate the different parameters can be found in [173].

In this article, we focus on the results of FS3 (competitive conventional power plants and untapped renewable potential) and FS4/5[6] (energy transition without the support of the population/cross-sectoral energy transition). The FS3 is characterized by low economic growth and low growth of the capital stock as well as no structural change in the agriculture and service sectors. Further, we assume historical autonomous energy efficiency improvements (AEEI) rates, constant primary energy prices, and an 80% reduction of CO_2 emissions in NWG. In FS4/5, we assume high economic growth and high growth of the capital stock as well as structural change in the agriculture and service sectors. Further, we define optimistic AEEI rates, rising primary energy prices and 100% renewables in NWG's electricity sector.

For both FS, we investigate the influence of three different policy strategies related to renewable energy promotion in North-West Germany (NWG). These so-called alternatives (see Section 8.2) can be influenced by regional governments. The interdisciplinary NEDS consortium has identified three alternatives by distinguishing seven fields where the federal states' governments can influence climate and energy policy. Because most of those fields are highly specific and cannot be captured by a CGE model, we focus on the different technologies used for power generation in NWG and the related costs among the following three alternatives [173]:

- A1: Local power generation with flexible management; focus on onshore wind power and rooftop PVs (more decentralized)
- A2: Large-scale storage and power generation; focus on offshore wind power and ground-mounted PV systems (more centralized)
- A3: Baseline

---

[6] FS4 and FS5 are equivalent in the economic model implementation.

9.9 Life-Cycle Assessment for the Derivation of Environmental and Social Preference Scores

M. Dumeier

To assess the environmental and in parts the social sustainability of the electricity generation system introduced above, a Life-Cycle assessment (LCA) approach was applied. An LCA is a systematic analysis of the potential-environmental impacts of products or systems throughout their entire life cycle according to the ISO Norm 14040, which defines a framework for the execution and documentation of an LCA [190, 191, 192].

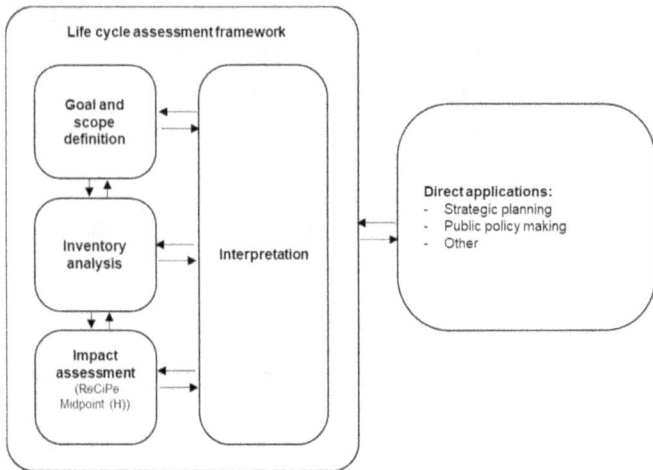

Figure 74: The Life-Cycle Assessment framework and its steps in the project [193]

In accordance with the ISO 14040, there are four stages to conduct an LCA (cf. Figure 74). First, the **goal and scope definition** provides an explanation of the intended application of the LCA. The main goal of the LCA in this case, is to measure the ecological and social impacts of different expansions of the renewable power generation systems in Lower-Saxony. Second, in the **inventory analysis** phase, the mass relations of all input- and output flows of the power generation system are collected and impact categories are defined. Third, the **impact assessment phase** aims to estimate the environmental impacts of the power generation systems concerning specific impact categories. The ReCiPe [194] model is used to calculate performance scores for impact categories of the power system. Lastly, the **interpretation phase** is used to combine the results of the impact assessment and inventory analysis phases and to compare the different configurations of the power system. In the following section, a more detailed explanation of the individual phases as executed in the project is presented.

The goal and scope of the LCA are largely derived from the overall targets of the project as described in Section 3.1. The overall goal is to compare multiple expansion pathways of a renewable power system regarding multiple criteria until 2050. Therefore, the goal of the LCA method is to identify and quantify ecological and social impacts of different configurations of the power system.

The product system in our study is the power generation system in Lower-Saxony encompassing the following power producers:

- Rooftop photovoltaic systems
- Ground-mounted photovoltaic systems
- Offshore wind turbines
- Onshore wind turbines
- Biogas power plants
- Hydropower power plants
- Lignite coal power plants
- Nuclear power plants
- Bituminous coal power plants
- Natural gas power plants
- Other (smaller cogeneration plants, waste and oil-fired power stations)

The product system also includes the required infrastructure for the transformation and transport of electricity, the high- and medium-voltage grid measured in km as calculated in Section 9.7 are incorporated accordingly. The functional unit is the reference to which all factors are normalized. It is based on the electrical energy (MWh_{el}) required to meet the demand of the end customers in the state of Lower Saxony in each of the reference years (2020, 2030, 2040, 2050).

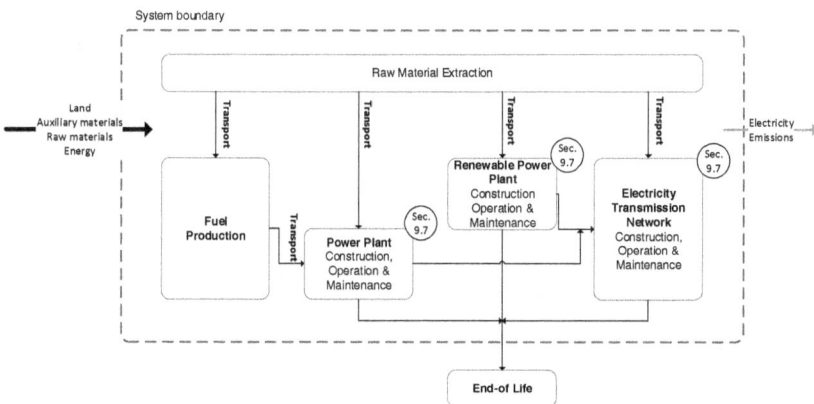

Figure 75: System Model of the considered electricity generation system

The technical system boundary is illustrated in Figure 75. The calculated emissions include both downstream activities, such as the construction of the power plants and upstream activities such as the transmission of this electricity. The geographic system boundary is derived from the location of the power producers and the purpose of the product system. Only power plants and transmission grid erected in the state of Lower Saxony are considered. The impact categories were selected in accordance with the results concerning the definition of sustainability described in Section 5.1 and by considering indicators used in related studies [195, 196].

Table 24: Description of impact indicators [197]

Impact Indicator	Unit	Impact	Social/ Environmental Sustainability
photochemical oxidant formation	Non Methane Volatile Organic Compounds ($NVMOC$)	Also referred to as "summer smog". Can have a variety of negative impacts on human health, ranging from respiratory symptoms to death.	Social
human toxicity	1,4-dichlorobenzene-equivalents ($1,4\,DC$-Eq)	The toxicity potential expressed in (1,4 DCB-eq). The assessments of the toxicity are based on the tolerable intake for humans.	Social
particulate matter formation	PM_{10}-Eq	Long-term exposure may lead to reduction in life expectancy due to cardio-pulmonary and lung cancer. [198]	Social
metal depletion	Iron equivalents Fe-Eq	Measures the extraction and consumption of metals that may thus not be available for future generations.	Environmental
fossil depletion	Oil equivalents oil-Eq	Measures the extraction and consumption of fossil fuels that may thus not be available for future generations.	Environmental
climate change	Carbon dioxide equivalents CO_2-Eq	Emissions of greenhouse gas emissions lead to increased global temperatures by increasing radiative forcing capacity	Environmental
terrestrial acidification	Sulfur dioxide equivalents SO_2-Eq	Measures substances that cause a change in the level of acidity in the soil, which can cause a shift in species occurrence	Environmental
freshwater eutrophication	Phosphorus equivalents P-Eq	Measures increase in nutrient levels in water bodies, which can ultimately lead to loss of species.	Environmental
terrestrial ecotoxicity	1,4-Dichlorobenzene 1,4- DCB -Eq	Measures the degree to which pollutants have an effect on land-dependent organisms. [199]	Environmental
agricultural land occupation	square meters per annum m^2a	Measures the amount of agricultural land occupied.	Environmental

In an LCA, midpoint and endpoint indicators can be used to measure impacts. Midpoint indicators are located between the stressor and endpoint. The endpoint indicators measure directly the factors valued by society. Midpoint indicators include factors such as human toxicity and particulate matter formation. Both of these factors can also have an effect on the endpoint indicator of human health. The drawback of the increased aggregation required by endpoint indicators is that additional information and uncertainty is introduced for instance through uncertain weights. In this study, midpoint indicators are used instead of endpoint indicators. These indicators are assigned to social and ecological sustainability dimensions, as described in Section 5.2. Table 24 provides an overview of the different impact indicators and the impacts that they model. The results for the midpoint indicators are discussed in detail in Section 10.5. An overview of all midpoint indicators is presented in the decision table of the multi-criteria decision problem (Table 35).

9.10 Coupling of Models on Macro-Level

C. Blaufuß, M. Dumeier, M. Hübler, H. Krause

On the macro-level, there are three major links. Between the model INES (Integrated Grid and Market Model) and the computable general equilibrium model, between the INES model and life cycle assessment, and between the optimized distribution grid planning model and life cycle assessment. Furthermore, a link exists between the optimized distribution grid planning model and the mosaik-connected models described in Section 9.4.

Between the INES [166] and the economic computable general equilibrium (CGE) model calibrated in the forward calibration process [173], we have established a soft link for exchanging relevant results. INES provides the electricity mix for the model years 2020, 2030, 2040 and 2050 and marginal electricity generation costs for conventional energy carriers. This information is used as an input in the CGE model. The CGE model is then solved for the CO_2 prices of the corresponding years given these electricity mixes, the CO_2 emissions targets and the other parameter settings of the scenarios (see Table 23 in Section 9.8). The CO_2 prices are then shared with INES. Furthermore, INES obtains the scenario-specific prices for primary energy carriers from the CGE forward calibration.

Between the life cycle assessment and the INES, a soft link exists. As is the case with the CGE model, INES provides the electricity mix for the model years 2020, 2030, 2040 and 2050. Along with the length of the high- and medium-voltage grid measured in *km* obtained from the optimized distribution grid planning model, these values are used to calculate the impact indicators as described in Section 9.9.

Finally, a link exists between the optimized distribution grid planning and with the mosaik-connected models from Section 9.4. The mosaik-connected models calculate the load and feed-in profiles for the MV/LV-interface. In this connection, the maximum value of the load and the feed-in serves as input data for the optimized distribution grid planning model and forms the important layout data. These data are received for every year of calculation.

10. Results of the Energy System Models

C. Blaufuß, M. Dumeier, M. Hübler, H. Krause, M. Nebel-Wenner

In this Section, the results for the scenario "Competitive conventional power plants and untapped renewable potential" and the three alternatives (see Section 8.2) are presented. The following sections describe in detail the results of the developed and applied models, as well as the life-cycle assessment (see Section 9). The results are evaluated with regard to the achievement of the objectives of the investigation.

10.1 Smart Grid Model

M. Nebel-Wenner

In this Section, the simulation results of the smart grid model ISAAC will be presented and discussed. A detailed analysis of the results regarding the multi-objective approach can be found in [165]. As described in Section 9.3, the model simulates a distributed optimization process for the scheduling of partly flexible domestic loads. Different simulations have been executed including varying parameter settings (see Section 9.5 for a description of the scenarios). In particular, the number of households that take part in the optimization varies between transition years and alternatives. As we assume an increasing number of smart meter over time, the strongest effects of the optimization can be seen in the transition year 2050. In the alternative 'decentral' we assume the highest willingness to take part in the optimization and hence, in the alternative-year combination 'decentral'-2050' the optimization effect is most prominent.

Figure 76 and Figure 77 show the simulation results regarding the development of the load of the interconnected buildings before and after optimization for the different simulated days and for the two grids for the alternative 'decentral' and the transition year 2050. It becomes apparent that the maximum grid usage can be significantly decreased due to the optimization process. This works specifically well when there is a peak of load after 8 pm. The time and load structure of the households indicate that such peaks stem from charging processes of electrical vehicles. If there is a high simultaneously in charging the electric vehicles, this can be smoothed out by shifting part of the charging processes toward later times.

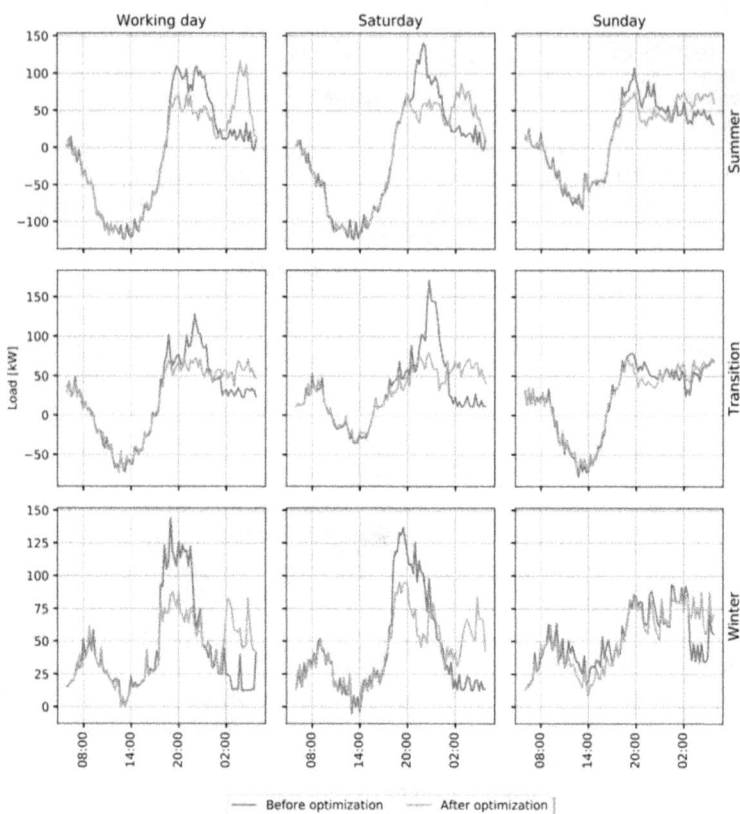

Figure 76: Simulation results for the VPP in the rural gird for the transition year 2050. 32 units were part of the optimization. Source: [165].

Table 25: Optimization effects for the VPP in the rural grid in the transition year 2050. The values describe the difference between simulation results before and after optimization. Source: [165].

Simulation Day	Peak load reduction	Market costs reduction in €	Behavioral adaptation costs
Working day Winter	56.8kW (39.4%)	9.3€ (3.7%)	0.86€
Working day Transition	56kW (43.5%)	5.6€ (3.6%)	1.90€
Working day Summer	5.36kW (4.4%)	7.11€ (11.2%)	1.26€
Saturday Winter	41.72kW (30.5%)	4.24€ (1.9%)	1.00€
Saturday Transition	92.42kW (54%)	4.94€ (2.9%)	1.55€
Saturday Summer	26.19kW (18.6%)	4.22€ (7%)	1.22€
Sunday Winter	5.47kW (5.9%)	4.51€ (1.8%)	1.42€
Sunday Transition	7.75kW (9.8%)	5.37€ (3.8%)	1.85€
Sunday Summer	32.21kW (29.9%)	5.78€ (4.6%)	1.84€

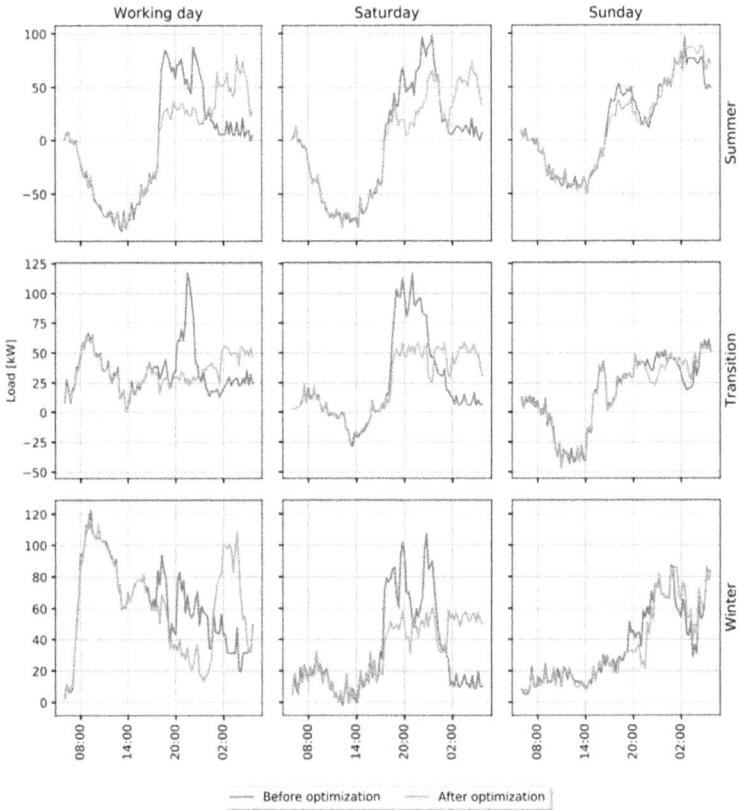

Figure 77: Simulation results for the VPP in the urban grid for the transition year 2050. 30 units were part of the optimization. Source: [165].

Table 26: Optimization effects for the VPP in the urban grid in the transition year 2050. The values describe the difference between simulation results before and after optimization. Source: [165].

Simulation Day	Peak load reduction	Market costs reduction in €	Behavioral adaptation costs
Working day Winter	8.86kW (7.2%)	5.54€ (1.5%)	0.29€
Working day Transition	55.84kW (47.4%)	5.33€ (2.8%)	1.22€
Working day Summer	6.87kW (7.8%)	6.53€ (11.6%)	0.85€
Saturday Winter	47.88kW (44.3%)	3.37€ (2%)	0.76€
Saturday Transition	57.96kW (49.5%)	3.83€ (2.9%)	1.70€
Saturday Summer	16.94kW (17.2%)	4.54€ (8.4%)	1.24€
Sunday Winter	1.15kW (1.3%)	2.83€ (1.6%)	0.75€
Sunday Transition	5.77kW (9.4%)	1.48€ (1.5%)	1.23€
Sunday Summer	8.76kW (9%)	2.54€ (2.2%)	2.24€

The figures additionally indicate that electricity costs are reduced. In all simulations, a significant amount of load is shifted toward the times with low electricity costs (between 0:00 and 5:00). Table 25 and Table 26 show that this indication is true, as in all simulations the electricity costs decrease after optimization.

However, the underlying flexibilities of the households are not sufficient to successfully match generation and consumption of electricity. That becomes apparent in simulations, in which a negative peak appears due to excessive feed-in of PV-systems. The feed-in peak in such simulations is hardly reduced after optimization. The reason for that is that there is not enough flexibility to shift a significant amount load to such periods. This is due to the fact, that most electric vehicles are not present during the day (and hence they cannot be charged) and that the user-driven flexibilities are of a rather small amount and costly to shift.

In the smart grid model, a specific emphasis has been put on the exploitation of the flexibility space of battery storages. However, Figure 78 indicates, that storages do not have a great impact on the optimization result. The main reason for that is that the battery storages fulfill a primary use case (optimization of self-consumption) including an 'immediate charging' strategy, which does not leave enough flexibility to significantly influence the optimization result. This is especially true for days with a high amount of feed-in of PV-systems, such it is the case in the simulated summer days in 2050. In Figure 78, it becomes apparent, that if the storages did not charge immediately, when there is excessive feed-in, but would charge during periods of maximum feed-in, the feed-in peak could be reduced without affecting the primary use-case of the battery storages. The same holds for the periods of discharging: if the batteries discharged mainly during periods of high demand and not as soon as demand exceeds the supply, the peak load would be reduced without affecting the goal of optimizing self-consumption.

Overall, the results show that a distributed multi-criteria optimization process can be successfully implemented at the household level. We showed that the given flexibilities of smart buildings can be used to pursue different targets. In order to integrate the different objectives into one target function, we chose a monetarization approach. This seems justifiable, given that the use case under investigation describes a rather bottom-up approach, in which interconnected smart buildings optimize their electricity consumption in order to get a monetary benefit. However, converting any criteria to a monetary scale is not without limitations. In our case, we faced the challenge to convert behavioral efforts into a cost value. The approach that was used in this project is presented in Section 9.1.3. However, the given approach generates data that should not be interpreted in terms of their absolute money values as they are just used for scaling purposes.

Regarding the simulated load curves, it becomes apparent that, if the number of electric vehicles increases massively, the corresponding charging processes play an important role in a future electricity grid. Our simulations confirm that the concurrent charging of electrical vehicles produces high load peaks, if uncontrolled.

We simulated nine different days including different weekdays and seasons. However, simulations remained in a 24 hours frame, which leads to limited representation of flexibility. This can be seen when looking at the flexibility of electrical heat pumps. Such peaks can be observed in some winter days in the morning (e.g., the working day in winter in Figure 78 In reality, heat pumps have certain flexibility to shift part of their needed electricity consumption toward earlier times. However, as all our simulations start at 6:00, this flexibility is not provided in the simulations. In future work, simulations of longer periods must be executed to see, if load peaks can be reduced further by using the flexibility that is beyond the 24 hours frame with the starting point of 06:00, which is used in this setting.

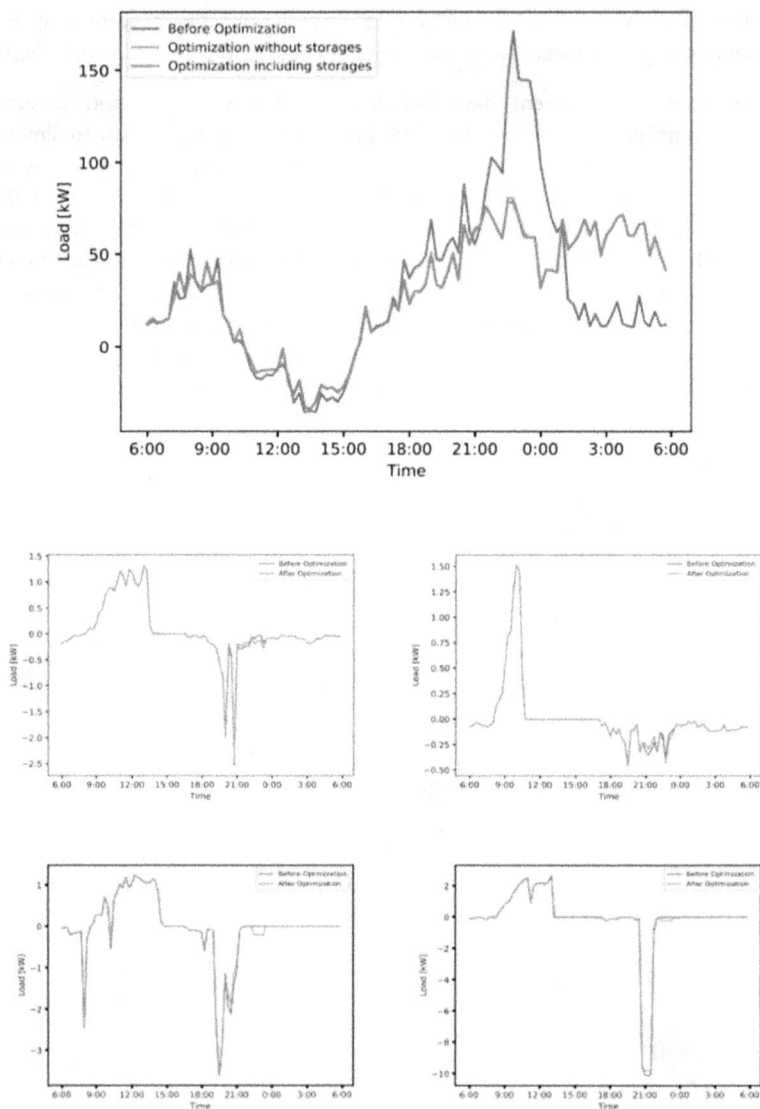

Figure 78: Simulation results for the storage systems. Simulation day: Saturday Transition 2050. Grid type: rural. The large plot displays the overall optimization effect with and without storages. In the small plots, schedules of four of the storage systems within the VPP are displayed exemplarily.

10.2 Results of the Integrated Grid and Market Model

C. Blaufuß

The results of the Integrated Grid and Market Model for the target states and the transition paths are used to assess the respective grid state. The target states and their transmission paths are formed by different alternatives, denote by Central, Medium and Local. Furthermore, the alternatives consist of different load assumption and alternating storage capacities, which are shown in section 6 and 8. The criteria, described in section 0, are used to assess the sustainability of the target conditions as well as their transmission paths consisting of the grid efficiency, the electricity mix and the CO_2-emission reduction. An example of the Energy mix of the target state in the year 2050 is given in Figure 79. The electrical mix is dominated by wind and solar energy in all different alternatives of the scenario. Furthermore, conventional, gas-fueled power plants have a percentage of five percent and operating while wind and solar plants can not to feed-in in the cause of wind and sun lacks.

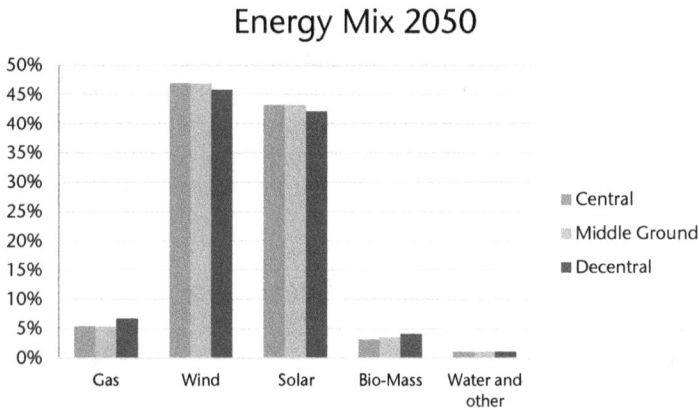

Figure 79: Example of the electricity mix of the year 2050

Additionally, the trends of the CO2-emission reduction and the grid efficiency are represented in Figure 81 and Figure 82 to benchmark the specific transmission paths.

The comparison of the results in Figure 80 shows in a similar trend of the grid efficiencies in the variations of Central and Middle Ground. Only the trend in the Decentral variation exhibits a deviating behaviour. The application of local storages with small capacities shows an improvement of the grid efficiency in the first years.

Grid Efficiency

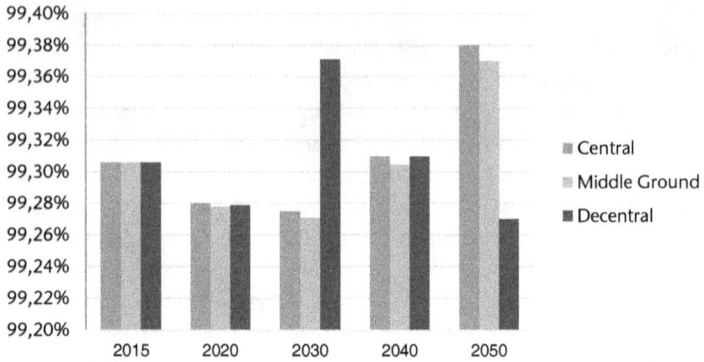

Figure 80: Trends of the grid efficiency

This deviation can be traced by the high number of remaining conventional power plants in the grid. The conventional power plants compensate inactive times of renewable power plants causing by lacks of wind and sun. In result, the need for the storing of electrical energy is reduced and only storages are required to intercept feed-in peaks of renewable power plants. Afterward the grid efficiency of the Decentral variation decreases and shows the worst state of the alternatives at the year 2050. This trend can be explained by the influence of renewable power plants, the missing of conventional power plants and the requirement of large storage systems.

CO_2-Emission Reduction

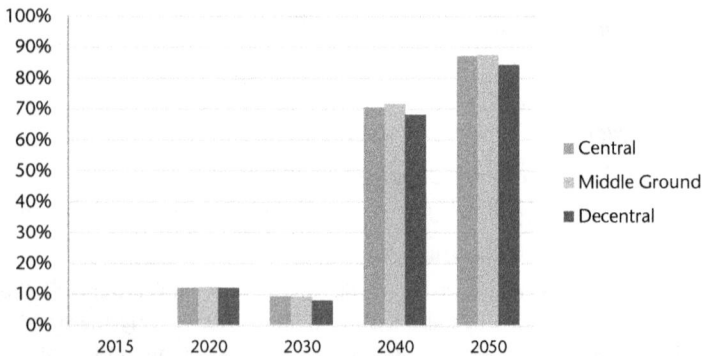

Figure 81: Trends of the CO_2-emission reduction

Figure 81 shows a similar trend of the CO_2-emission reduction over all variations due to the equal high integration of renewable power plants. The worst variation is represented by the local variation. This result can be traced back to the inactive times of renewable power plants causing by lacks of wind and sun, which cannot adequately compensated by the stored electrical energy. In result, gas-fired plants compensate for the lack of energy in the grid and produce CO_2 as a by-product. Furthermore, an interesting effect is visible and is shown in a regressive behaviour of the CO_2-emission reduction in the year 2030, due to the exit of nuclear energy in previous years.

10.3 Results of the Grid Planning Algorithm

C. Blaufuß

The necessity of grid expansions is representative by calculated for one exemplary high- and one medium-voltage grid. Furthermore, the simulations are executed for the target states as well as the states of the transition paths. Figure 82 shows an example of the grid extension of a medium-voltage grid, starting from the year 2015 and ending with the target state of the year 2050.

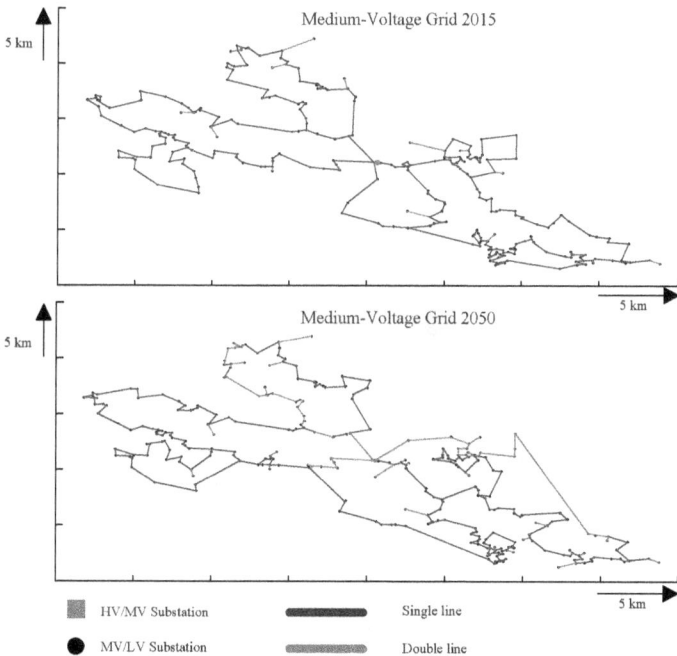

Figure 82: Example of the grid extension of the MV Grid

The load and feed-in assumptions as well as grid connection points for the accommodation of the additional loads and feed-ins increase and resulting in a need of grid extensions.

As a result, the number of MV-nodes builds up from 280 to 307 with progress to the year 2050. It copes with the additional stress, on the one hand, by the reconstruction of the grid topology, shown in the bottom part of the grid at the year 2050. On the other hand, by executing grid reinforcements, shown by the red connection lines in Figure 82.

Other results are shown in Figure 83 and Figure 84 and represent a comparison of system lengths in the HV and MV Grid for the target state and the transition path. It shows an increase of the system length in both levels, regarding the system length of the year 2015. The system length accomplished their maximum at the year 2040 in two alternatives of the scenario (decentral and central). The increase of the system length can be explained by the additional stress for the grid due to the increased load and feed-in assumption as well as the building of new grid connection points. Furthermore, it is difficult to give a statement about the system lengths development of the respective transmission path because they are additionally dependent on the age of the existing grid and the loads in the respective years.

System Length HV-Grid

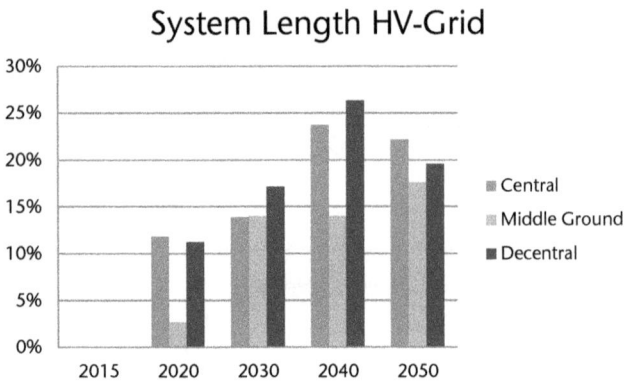

Figure 83: System length of the HV Grid

System Length MV-Grid

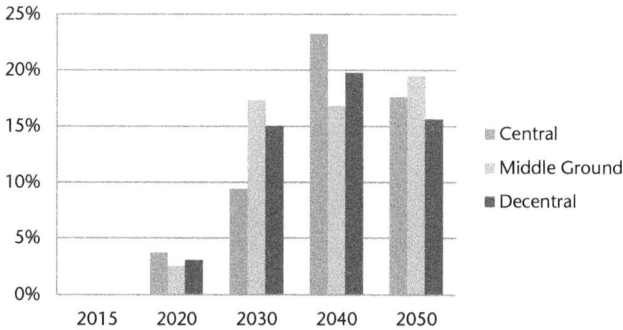

Figure 84: System length of the MV Grid

Additionally, the annual costs are considered as a second criterion for evaluation. This comparison refers to the annual grid cost of the starting state and is shown in Figure 85 for the high-voltage grid as well as in Figure 85 for the medium-voltage grid. No costs or maybe negative costs can be interpreted as a low or none-existing grid extension, due to the depreciation of the transmission elements and lose their worth. These effects can be recognized in the high-voltage grid costs of the year 2020 and 2040, caused by the low gradient of the load and feed-in values.

Annual Cost of the HV-Grid

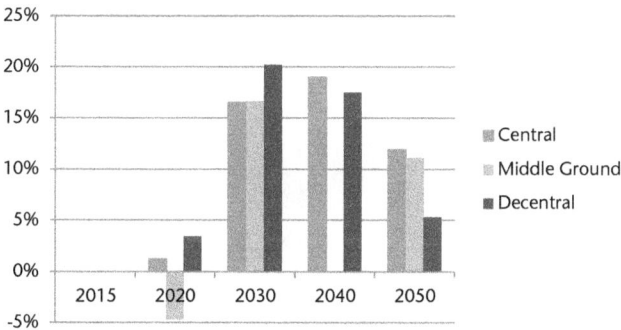

Figure 85: Annual cost of the HV Grid

Annual Cost of the MV-Grid

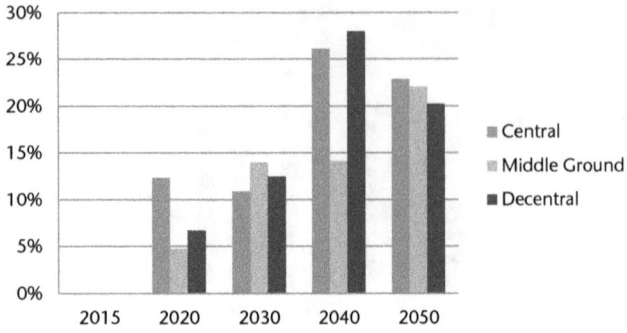

Figure 86: Annual cost of the MV Grid

10.4 Macroeconomic Developments under Different Energy Policies

H. Krause, M. Hübler

In this Section, we discuss the results of the simulations developed in this project to analyze the influence of energy policies on Lower Saxony's (LSX) or alternatively Northwest Germany's (NWG) economies (see Section 9.8 for model details). To compare the results, the following simulations are reviewed: (i) baseline scenario for Lower Saxony (LSX) under the existing European Union Emissions Trading System (EU ETS) and (ii) future scenarios (FS) that describe the most likely economic developments between the years 2011 and 2050. To evaluate the effects of different policies under the different scenarios, we subsequently introduce three alternatives (A1, A2, and A3) that are introduced into the future scenarios. This allows us to compare the effects of the policies under different economic development conditions.

The results of the analysis of the baseline scenario for 2011 show that under the European Union emissions trading system (EU ETS), LSX can reduce CO_2 emissions by 26.5% [176] with climate policy compared to no policy, which is about three times as much as the percentage changes in the rest of Germany (ROG_1) and almost five times as much as in the rest of the European Union member states. The main reason for this is that LSX has a large capacity to expand wind power, which is a relatively cheap renewable energy source [176]. This lowers the CO_2 reduction costs of this region.

If policy-makers add additional renewable energy support to the EU ETS (ETS/RES), renewable energies will expand significantly in LSX such that particularly wind energy will be used abundantly (see Figure 87 and Figure 88). The use of hydropower and geothermal energy, on the opposite, will be lower than under the EU ETS alone or under the baseline situation. Because of the higher usage of electricity under renewable energy support, the CO_2 reduction in LSX is lower than under the EU ETS alone, but the renewable electricity share reaches 94% in 2011. Because of the additional subsidies, the welfare loss and thus the social cost of the energy transition in LSX are higher under the combination of the EU ETS and renewable energy support (-0.27% of total consumption) than under the EU ETS alone (-0.04%).

In summary, LSX's CO_2 emission reductions are larger while the welfare losses are smaller than in the rest of Germany. One reason is that LSX has a comparative advantage in terms of CO_2 abatement costs due to its large wind power potential. Renewable energy support in other European countries, however, reduces this comparative advantage [176]. The results also show that it is relevant to disaggregate the electricity sector for a meaningful analysis of climate policies.

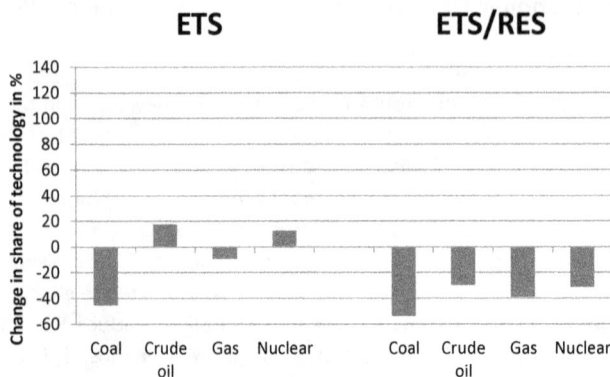

Figure 87: Policy-induced change of conventional energy carriers in the electricity mix of Lower Saxony compared to no policy. Notes: ETS = European Union's emissions trading system; ETS/RES = EU ETS plus renewable energy support by Lower Saxony (Source: [176], Appendix M, Figure M1, p. 23)

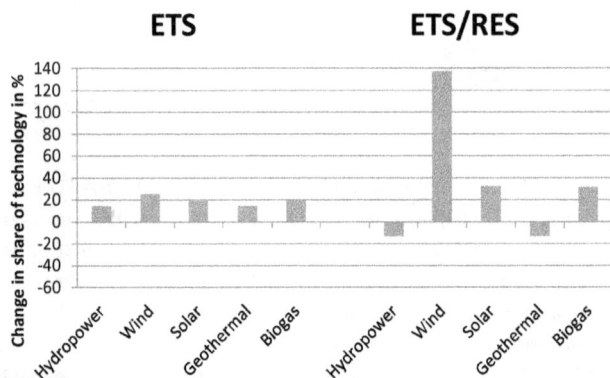

Figure 88: Policy-induced change of renewable energy carriers in the electricity mix of Lower Saxony compared to no policy. Notes: ETS = European Union's emissions trading system; ETS/RES = EU ETS plus renewable energy support by Lower Saxony (Source: [176], Appendix M, Figure M1, p. 23)

Next to the effect of climate policies on the baseline scenario, we investigate the effect of two different trade policies and the interaction between trade policies and climate policies (see [176] and Section 9.8). The results show that both LSX and ROG$_1$ profit from a reduction in trade costs (see Table 27 and Table 28). The overall welfare measured as real consumption increases by 0.42% in LSX [176], if all tariffs are abolished. Reducing the iceberg trade costs yields even twice this welfare gain. In ROG$_1$, the percentage increase in real consumption is smaller than in LSX, indicating stronger importance of trade for LSX's economy as compared to the aggregated rest of Germany.

The reduction of trade costs decreases total CO_2 emissions slightly in both regions, LSX and ROG_1, mainly dedicated to the slight decrease in gross output in both free trade scenarios. The share of renewable energies in LSX's and ROG_1's electricity mix increases slightly through the reduction of trade costs and the CO_2 price increases by around 35% in both regions [176].

Table 27: Results of trade scenarios and interactions with EU ETS for LSX

Indictator	noTarrifs	noTarrifs/ETS	lessIceberg	lessIceberg/ETS
Real consumption	0.42	0.40	0.85	0.83
Gross output	-0.06	-0.40	-0.41	-0.75
Total CO_2 emissions	-7.41	-29.65	-6.94	-29.25
Share of renewables	2.40	23.51	0.69	21.83
CO_2 price	35.13	190.12	33.71	186.99

Note: All indicators are expressed as percentage changes compared to the baseline scenario with none of the investigated policies; Source: [176], Table 6, p. 19 and Table 8, p. 23.

If we consider the interaction effects between free trade and the European Union's emissions trading scheme (EU ETS), we can see a strong increase of the CO_2 price to about 190% as compared to a scenario with neither of the two policies (see Table 27 and Table 28). This leads to a rather strong increase of the renewable energy share in electricity generation in LSX and ROG_1 of more than 20% accompanied by a decrease in CO_2 emissions, which is much stronger as under the free trade scenario alone. While the increase in renewable energy share is similar in LSX and ROG_1, LSX is able to reduce its CO_2 emissions more effectively by substituting fossil energy carriers by renewable energy. One reason for this is the already discussed low abatement costs in LSX thanks to the potential for wind energy in this region.

Table 28: Results of trade scenarios and interactions with EU ETS for ROG_1

Indictator	noTarrifs	noTarrifs/ETS	lessIceberg	lessIceberg/ETS
Real consumption	0.23	0.18	0.66	0.61
Gross output	-0.51	-0.63	-0.78	-0.9
Total CO_2 emissions	-2.77	-12.35	-2.66	-12.18
Share of renewables	3.73	24.47	2.6	23.19
CO_2 price	35.13	190.12	33.71	186.99

Note: All indicators are expressed as percentage changes compared to the baseline scenario with none of the investigated policies; Source: [176], Table 6, p. 19 and Table 8, p. 23.

Interestingly, the high CO_2 price in the combined scenarios does not cause a substantial decline in gross output and does not decrease the welfare gains observed in the scenarios with free trade only either. This shows that it is important to consider both, trade policy and climate policy, together. Well-implemented trade policies have the potential to complement the CO_2 reduction effects of climate policies and help to mitigate short-term negative economic effects, especially in an economy like LSX with the capacity to reduce CO_2 at relatively low costs in the electricity sector.

Conclusively, the welfare gains of removing non-tariff trade barriers are much higher than the equivalent removal of tariffs [176]. This outcome corresponds with recent regional trade agreements that go beyond the removal of tariffs. Both measures slightly increase CO_2 emissions on a global scale because of an increased trade volume and thus increase in production. In LSX, however, both trade measures reduce CO_2 emissions, which renders the fulfillment of the EU emissions targets less costly. Thus, wise integrated and internationally coordinated climate and trade policy design can create additional benefits for climate protection and regional welfare [176]. These results contradict recent US policies fostering a renaissance of protectionism and ignorance of climate change.

While the EU ETS induces a moderate reduction of conventional energies in electricity generation and a moderate expansion of renewable energies, additional renewable energy support induces a strong expansion of wind power in LSX [176]. This renewable energy support policy, however, comes at an additional social cost. Furthermore, the effectiveness of Lower Saxony's renewable energy support based on Lower Saxony's comparative advantage in deploying wind power is mitigated by corresponding renewable energy support abroad [176].

In the second step of our analysis, we implement future scenarios and analyze the effect of different policy alternatives on different economic indicators [173]. The spotlight is on the region of Northwest Germany (NWG), which comprises the federal states Lower Saxony, Hamburg and Bremen and the time frame of the model is until 2050.

Under the realistic medium scenario FS3, the NWG economy experiences a slight increase in per capita GDP and capital stocks until 2050 (see Table 29). The productivity stagnates until 2020 and then increases by a bit more than 10% relative to 2011 until 2050. The increase of per capita GDP is related to the trade linkages between NWG, the rest of Germany (ROG$_2$) and the European Union in addition to own productivity gains.

Table 29: Results FS3 and FS4/5 for NWG

Indicator	FDS	2011	2020	2030	2040	2050
GDP per capita in 1000 €	FDS3	33.03	34.98	37.27	39.79	42.54
	FDS4/5	33.03	37.23	42.53	48.67	55.77
Weighted average import share of primary energy carriers in %	FDS3	18.66	12.20	12.96	9.17	3.87
	FDS4/5	18.66	13.36	12.77	3.51	0.00
Expenditure share of electricity in final consumption in %	FDS3	1.07	1.12	1.05	0.97	0.98
	FDS4/5	1.07	1.03	0.97	0.89	0.84
Wage-to-capital income ratio	FDS3	1.01	0.99	0.99	0.99	1.00
	FDS4/5	1.01	0.99	0.98	0.97	0.97
Total CO$_2$ emissions relative to 2011 in %	FDS3	100.00	77.63	59.83	44.09	27.23
	FDS4/5	100.00	84.35	61.38	44.69	28.43

Source: [173], Table 3, p. 26 and Table 4, p. 31 and further results of [173].

In FS3, total CO$_2$ emissions decline substantially until 2050 with -72.7% compared to 2011 in NWG and -76.5% in ROG$_2$ (see Table 29 and Table 30). For entire Germany, CO$_2$ emissions in the electricity sector decline to about 162 Mt, equivalent to 13.5% of the CO$_2$ emissions in 1990 [173]. Thus the goal of the German government to produce only 20% of the CO$_2$ emissions in 1990 by 2050 is reached. The reductions are higher in NWG until 2040 compared to ROG$_2$, suggesting lower CO$_2$ abatement costs in this region. This goes along with the results of the base model for 2011 described above.

The share of renewable energy in electricity generation reaches around 90% in both regions by 2050 and is larger in LSX than in ROG$_2$. Because of the increasing share of renewable energies in electricity generation, the import share of primary energy carriers declines in both regions, NWG and ROG2, to a fourth and a fifth of 2020 values, respectively. Thus, the EU ETS helps to reduce the dependency of NWG and ROG2 on external energy supply in the form of coal, gas, and oil.

Until 2040, the prices for CO$_2$ emission allowances under the ETS increase only slightly as compared to 2011. This shows that with moderate economic growth and current levels of autonomous energy efficiency improvements the EU ETS goals can be fulfilled without substantially increasing the price for CO$_2$ emission allowances [173]. Towards 2050, when almost 90% of the energy stems from renewable resources, the price for CO$_2$ allowances increases strongly. This is also visible in the reduction of real consumption in NWG and ROG$_2$, which stays relatively small until 2040 and increases substantially in 2050. Thus, the marginal cost of converting conventional energy into renewable energy increases strongly when renewable energy share reaches more than 80% or even 100%. In reality, converting only a

185

fraction of the energy production towards renewable production can be reached by changing the energy supply as such, but shifting the entire system to renewable energy requires changes in the grid and storage infrastructure and the phase-out of gas power as a backup for renewables. This increases the CO_2 abatement cost more than proportionately to the CO_2 reduction.

Under FS4/5, per capita GDP and capital stock growth are much higher than under FS3, mainly because the absolute productivity increase is larger (see Table 30). CO_2 emissions decline to about the same extent as in FS3 and the share of renewable energies in electricity generation is slightly higher. It reaches almost 100% in NWG by 2050, thus reducing the import of primary energy carriers to zero.

Table 30: Results FS 3 and FS 4/5 for ROG

Indicator	FDS	2011	2020	2030	2040	2050
GDP per capita in 1000 €	FDS3	32.30	34.20	36.45	38.91	41.59
	FDS4/5	32.30	36.40	41.58	47.59	54.53
Weighted average import share of primary energy carriers in %	FDS3	27.46	31.78	25.24	16.99	5.58
	FDS4/5	27.46	33.13	23.78	14.15	4.29
Expenditure share of electricity in final consumption in %	FDS3	1.07	1.06	1.05	0.99	1.00
	FDS4/5	1.07	1.02	1.00	0.93	0.91
Wage-to-capital income ratio	FDS3	1.09	1.07	1.07	1.07	1.07
	FDS4/5	1.09	1.08	1.07	1.09	1.11
Total CO_2 emissions relative to 2011 in %	FDS3	100.00	99.49	68.89	46.32	23.45
	FDS4/5	100.00	96.72	62.51	41.88	22.09

Source: [173], Table 3, p. 26 and Table 4, p. 31.

In all years, the prices for CO_2 emission allowances are higher in FS4/5 as compared to FS3. This is mainly due to the high economic growth rates in this scenario, which translates to higher CO_2 emissions. However, the assumed greater autonomous energy efficiency improvements, structural changes toward the service sector in Germany and a high share of renewable energies in electricity generation in this scenario partially compensate for this effect.

Despite the higher CO_2 prices, the negative welfare effects of the EU ETS are smaller in FS4/5 as they are in FS3. This surprising result can be explained by the high levels of productivity and energy efficiency in FS4/5 in the European economies, making them less dependent on energy usage. Furthermore, generating 100% of electricity from renewable energies in NWG makes the electricity sector and depending industries resilient towards the introduction of stringent climate policies.

For the two future scenarios FS3 and FS4/5, we calculated the effects of two policy alternatives (A1 and A2) and compare them to the baseline alternative (A3) [173]. In the first alternative (A1), the electricity generation is shifted toward higher shares of onshore wind power and rooftop photovoltaics (PVs) compared to the baseline, while in the second alternative (A2), the electricity generation changes toward higher shares of offshore wind energy and ground-mounted PVs.

Rooftop PVs are more expensive per MW than ground-mounted PVs, whereas onshore wind power is cheaper than offshore wind power. While the overall effects of the two alternatives on GDP are small, the results show a slightly negative effect of A1 and a slightly positive of A2 [173]. The negative effect in A1 suggests that the relatively higher price for rooftop PVs is not offset by the relatively lower price for on-shore wind power (see Table 31).

Productivity effects of the two alternatives are only visible in NWG in individual years. However, since the economy-wide productivity is first rising and later falling, the overall effect of both alternatives on NWG's productivity is ambiguous.

Table 31: Results of the alternatives for FS3

Indicator	Re-gion	A1				A2			
		2020	2030	2040	2050	2020	2030	2040	2050
GDP per capita	NWG	0	-	-	-	0	+	+	+
	ROG	0	0	0	0	0	0	0	0
Regional absolute productivity	NWG	+	+	-	-	+	+	-	-
	ROG	+	+	+	-	+	+	+	-
CO2 emissions	NWG	-	+	-	-	-	+	-	-
	ROG	+	-	+	+	+	-	+	+
Share of renewables	NWG	+	+	+	+	+	+	+	+
	ROG	-	+	-	-	-	+	-	-
Import share of primary energy carriers	NWG	-	-	-	-	-	-	-	-
	ROG	0	-	-	+	0	-	-	+
Share of electricity in consumption expenditures	NWG	+	+	+	+	+	+	-	-
	ROG	-	+	0	0	0	+	-	-
Wage-to-capital income ratio	NWG	+	+	+	+	+	+	0	0
	ROG	0	0	0	0	0	0	0	0
Change in EU ETS price		-	+	-	-	-	+	-	-

Notes: All results indicated as qualitative changes from the baseline (A3); + increase, - decrease, o constant. Source: [173], Appendix E, Table E3, p. 56.

Under both alternatives, the share of renewables increases significantly in NWG as compared to the baseline. In reality, larger storage capacities for electricity are available in A1 and A2 and thus less gas power plants are needed to stabilize the power in the electricity grids compared to the baseline scenario [166].

With higher shares of renewables in power generation, imports of primary energy carriers decrease in both scenarios, reducing NWG's dependency on those resources. This reduces the CO_2 emissions for NWG. The price for CO_2 emission allowances of the EU ETS decreases slightly by A1 and A2.

While the influence of the two alternatives on NWG's and ROG_2's GDP is the same in FS3 and FS4/5, the influence on productivity is stronger in FS4/5 than in FS3 (see Table 32). Especially in 2050, productivity increases to almost 3% above baseline scenario levels in A1 and almost 2% in A2, respectively.

The share of renewables is positively influenced by the alternatives under FS4/5, but the effect is weaker than under FS3. The alternatives favoring renewable energies have less influence in FS4/5, because there are already higher autonomous energy efficiency improvements and higher primary energy prices, leading to 100% renewable energy in NWG's electricity generation. Like under FS3, we can see a decrease in imports of primary energy carriers and a slight decrease in the CO_2 price under A1 and A2 for FS4/5.

Table 32: Results of the alternatives for FS4/5

Indicator	Re-gion	A1				A2			
		2020	2030	2040	2050	2020	2030	2040	2050
GDP per capita	NWG	0	-	-	-	0	+	+	+
	ROG	0	0	0	0	0	0	0	0
Regional absolute productivity	NWG	-	-	+	+	-	-	-	+
	ROG	0	+	+	+	0	+	-	-
CO2 emissions	NWG	-	-	-	0	-	-	+	+
	ROG	+	+	-	+	+	+	+	-
Share of renewables	NWG	+	+	+	0	-	+	-	0
	ROG	-	-	+	+	-	-	-	+
Import share of primary energy carriers	NWG	-	-	-		-	-	+	
	ROG	0	-	-	-	+	-	+	+
Share of electricity in consumption expenditures	NWG	-	+	+	+	-	+	-	-
	ROG	-	-	+	+	-	-	-	+
Wage-to-capital income ratio	NWG	-	+	+	-	-	+	-	-
	ROG	0	-	-	+	0	-	+	+
Change in EU ETS price		-	-	-	+	-	-	+	+

Notes: All results indicated as qualitative changes from the baseline (A3); + increase, - decrease, o constant. Source: [173], Appendix E, Table E4, p. 57.

Conclusively, the future scenario analysis indicates that in small open economies like Northwest Germany, international trade enhances economic growth, but domestic technological progress is required to achieve higher growth rates [173]. While economic growth creates additional CO_2 emissions, higher autonomous energy efficiency improvements and structural change toward less emissions-intensive sectors such as services can compensate this increase in CO_2 emissions.

The EU ETS effectively reduces CO_2 emissions in Europe at relatively low prices for CO_2 emission allowances across the sectors covered by the scheme. The overall costs in terms of GDP and consumption losses are small (far below 1%) for NWG and ROG unless emissions reductions of 80% are imposed. However, if NWG's energy transition fails, costs for climate policy will increase drastically when the reduction target approaches 80% towards 2050 [173].

The CO_2 abatement costs in sectors not covered by the EU ETS, however, are (from 2030 on threefold) higher [173]. Hence, a uniform CO_2 price covering all sectors including agriculture, transportation, etc. could significantly reduce the CO_2 abatement costs. Currently, only power plants, energy-intensive industries, and inner-European civil aviation are included in the ETS with a uniform CO_2 price. The rest of the sectors with about 60% of the EU's greenhouse gas emissions are currently not covered by the scheme [200].

The analysis of the policy alternatives shows that regional support for specific wind and solar power technologies increases the share of renewable energy carriers and decreases the import share of primary energy carriers [173]. Regional support for large-scale storage and large-scale renewable power generation with a focus on offshore wind power and ground-mounted PVs creates slightly positive effects on GDP and households consumption in Northwest Germany [173].

However, we need to keep in mind that the future scenarios are based on a set of assumptions about economic growth, structural and technological change, and climate policy. Crucial model parameters, particularly elasticities of substitution, are subject to uncertainty, see the robustness check and sensitivity analysis in [176]. Thus, the results need to be interpreted with caution and under the reflection of these assumptions and uncertainties.

10.5 Life-Cycle Assessment

M. Dumeier

To identify and quantify the environmental impacts of electricity generation in Lower Saxony until 2050 selected LCA impact categories have been calculated for the three alternatives previously defined:

- A1: Local power generation with flexible energy demand; focus on onshore wind power and rooftop PV ("decentral")
- A2: Large-scale storage and power generation; focus on offshore wind power and ground-mounted PV ("central")
- A3: Middle ground: a mix of all generation and storage technologies ("medium")

Table 33 shows the life cycle inventories for the three alternatives in the individual assessment periods. The energy production is expressed in TWh of electricity generated through different energy sources in the considered alternative and assessment year. The inventory of the grid is converted to the functional unit, described in Section 9.9. The total grid kilometers are consequently divided by the delivered electricity over the entire grid lifetime, assumed to be 40 years. The overall decrease in km per TWh can be attributed to the increased power production and only a minor increase in required grid kilometers. While the overall production increases up to the year 2050 by 56.4% [A1] – 59.9% [A3] in comparison to 2020 the total grid kilometers increase by 11.2% [A1] – 16.1% [A3] in the same timeframe.

Table 33: Life cycle inventories for the different energy sources and the transmission grid

		2020			2030			2040			2050		
	Source	S3 A1	S3 A2	S3 A2	S3 A1	S3 A2	S3 A2	S3 A1	S3 A2	S3 A2	S3 A1	S3 A2	S3 A2
PV	Rooftop [TWh]	7.692	6.569	7.131	16.595	11.182	13.887	23.227	10.359	16.790	26.903	4.100	15.838
	Ground-mounted [TWh]	2.306	3.429	2.867	7.495	12.908	10.203	14.955	27.823	21.392	24.071	48.174	36.436
Wind	Onshore [TWh]	16.807	16.407	16.609	25.476	24.135	24.808	32.198	29.383	30.792	36.057	32.157	34.570
	Offshore [TWh]	2.220	2.625	2.422	6.132	7.474	6.801	11.987	14.799	13.389	19.295	24.606	22.189
	Biogas [TWh]	8.478	8.598	8.560	4.715	8.577	8.476	6.533	6.728	7.103	5.045	3.916	4.250
	Hydropower [TWh]	0.935	0.935	0.935	0.756	0.935	0.935	0.936	0.936	0.935	0.913	0.936	0.936
Non-renewable	Bituminous coal [TWh]	19.611	19.675	19.598	16.822	19.563	19.622	1.865	1.472	1.339	0.310	0.281	0.254
	Lignite coal [TWh]	0.000	0.000	0.000	0.000	0.000	0.000	0.000	0.000	0.000	0.000	0.000	0.000
	Nuclear power [TWh]	16.327	16.551	16.551	0.000	0.000	0.000	0.000	0.000	0.000	0.000	0.000	0.000
	Natural gas [TWh]	2.910	2.769	2.843	9.748	5.369	5.316	13.895	13.313	13.137	8.175	6.612	6.538
	Waste [TWh]	0.000	0.000	0.000	0.194	0.000	0.000	0.070	0.065	0.052	0.045	0.038	0.037
Grid	Medium voltage [km /TWh]	50.233	50.375	49.821	49.186	45.728	49.081	42.644	44.216	41.903	35.131	36.643	37.146
	High voltage [km /TWh]	13.319	13.344	12.255	12.309	11.692	11.718	11.057	10.906	10.048	8.928	9.050	8.982

To illustrate the changes in the power system, we show selected results in more detail for the first (2020) and last (2050) assessment periods for the climate change impact factor. For a comprehensive overview of all considered impact factors for all of the assessment years, see Table 35.

Figure 89 shows the climate change related emissions for the year 2020. The results are shown for three sources for each alternative, which in together represent the total emissions. The electricity produced from fossil sources (50.3% of the total electricity production) creates approximately 88% of the climate change emissions. For renewable energy sources, electricity produced by biogas causes 59% of emissions (6% of total emissions). Around 0.7% (2.48-2.57kg CO_2-eq/MWh) of the total climate change emissions can be attributed to the transmission grid.

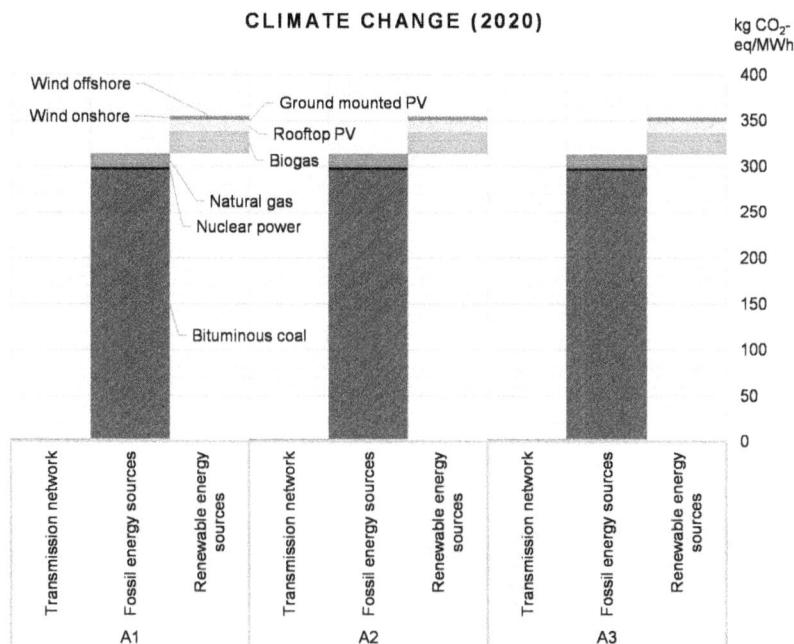

Figure 89: Climate Change Emissions in the year 2020 by source and alternative

In the year 2020, the energy mix, i.e., the contribution of the energy sources to the total electricity production is very similar among the three alternatives. This is also reflected by the emissions calculated for these alternatives. However, among the impact indicators, the contribution of the different energy sources varies. The transmission grid is responsible for 29% of the metal depletion in every alternative. This is the highest contribution by the transmission grid to any impact indicator. The second highest contribution of the transmission grid is to the agricultural land occupation indicator (6.12% [A3] – 6.21% [A1] of total land occupation). It is lowest for the climate change indicator where it contributes 0.70% in alternative 3 and 0.72% in alternative 1 of total climate change emissions. Fossil energy sources have the highest contribution to the freshwater eutrophication impact indicator (91.18% [A3] – 91.2% [A2]). The lowest contribution of this energy source is to the terrestrial ecotoxicity indicator (5.75% [A3] – 5.78% [A1]). For this indicator, the emissions are largely caused by the technologies utilizing renewable energy sources (91.5% [A3] – 91.21% [A1]). The majority of the terrestrial ecotoxicity can be attributed to biogas (61.65% [A1] – 62.3% [A2]). The contribution of renewable energy sources is also high for the particulate matter formation indicator (62.3% [A1] – 62.72% [A2]). For this indicator, biogas causes approximately 49% in every alternative. For the metal depletion indicator, Wind Onshore and rooftop PV systems have a minimum contribution of 39.2% for alternative A2 and a maximum contribution of 42.97% for alternative a1. For this indicator, the contribution of the transmission grid is highest (29.20% [A3] - 29.69% [A2])

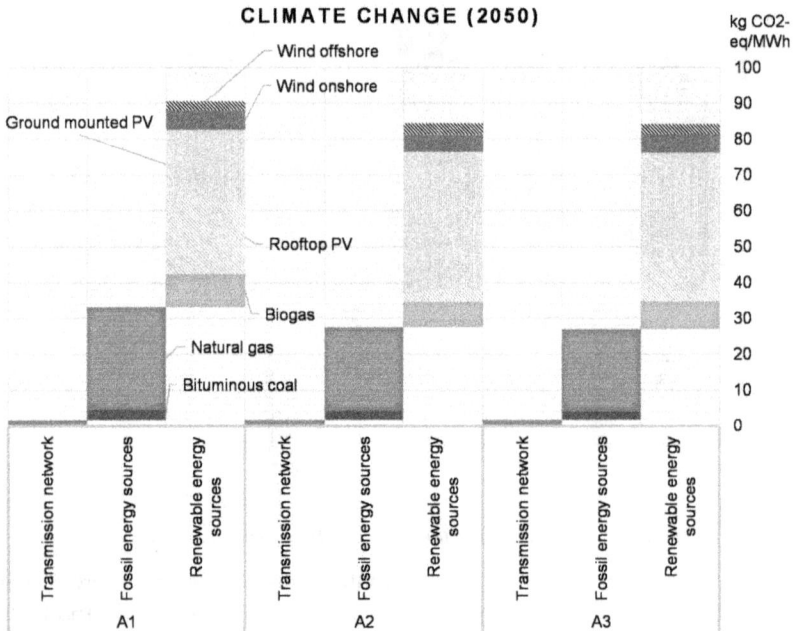

Figure 90: Climate Change Emissions in the year 2050 by source and alternative

Figure 90 shows climate change emissions for the year 2050. Overall, the emissions are reduced by 74.42% [A1] - 76.10% [A3], compared to the values in 2020. Power plants relying on fossil fuels are still responsible for 30.09% in alternative A3. The contribution of technologies relying on these energy sources to the over-all power production is lowest. Accordingly, the contribution is highest in alternative A1 (34.90%) which is also the alternative with the highest energy production by non-renewable power plants. In comparison to 2020, the highest reduction over all impact categories is in the freshwater eutrophication, which is reduced by 90.90% [A1] - 91.96% [A2]. Agricultural land occupation shows the smallest reduction (29.70% [A1] - 36.51% [A2]). There are also two impact categories that are increased compared to 2020, metal depletion by 37.79% [A2] - 46.74% [A1] and terrestrial ecotoxicity by 17.78% [A2] - 22.05% [A1].

11. Evaluation of Transition Paths

M. Dumeier, T. Witt, J. Geldermann

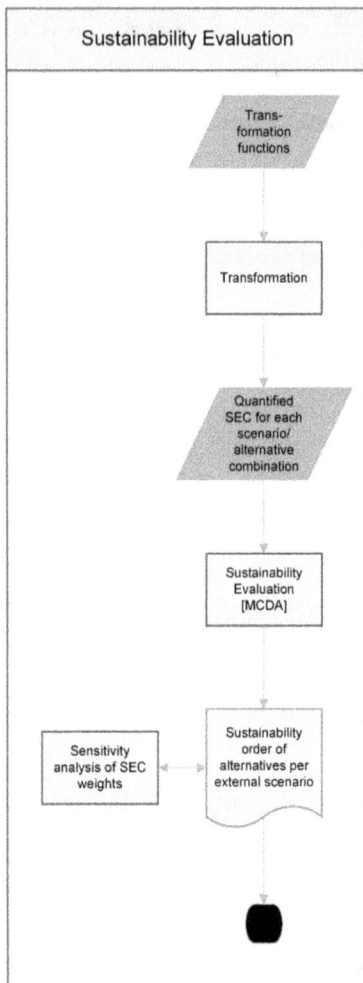

Figure 91: Excerpt from the PDES for evaluating the sustainability of alternatives in external scenarios

One of the central goals in the project is to develop and evaluate paths toward a sustainable energy supply system. To recapitulate the basis for such an evaluation: In Section 0, multiple sustainability criteria are defined for the evaluation of transition paths of Lower Saxony's power supply system. Given the future scenarios and transition paths (described in Sections 6 and 8), each transition path has been quantitatively modeled and simulated (see Section 9). Finally, selected results of this quantitative modeling, the so-called performance scores $x_{ij,t}$, i.e., the performances of all alternatives j on all sustainability evaluation criteria i in all years of the transition paths t, are forwarded to the evaluation via transformation functions (see Figure 91). To aggregate the performance scores of the 18 criteria obtained from the simulation and optimization models, we use a Multi-Criteria Decision Analysis (MCDA) method, namely a modified PROMETHEE approach [201].

The aim of applying Multi-Criteria Decision Analysis (MCDA) methods is to support the decisions of one or more decision makers in evaluating different alternatives regarding multiple often conflicting criteria given in different units of measurement [54]. The method also structures the decision problem and therefore allows for more informed and transparent decisions. The evaluation of transition paths reaching over decades is a highly complex decision problem with multiple conflicting criteria.

Textbook examples for MCDA describe the choice of a car or a supermarket location, with tangible alternatives and specific decision makers. In this case, the decision problem is not as [202] well defined. The method is applied to make explicit which factors drive the choice for a certain energy system configuration. In particular, we evaluate the transition paths consisting of alternative configurations of the power generation system in the state of Lower Saxony (alternatives) as defined in Section 8.3. The overall aim of the decision support process is to assess the sustainability of the transition paths according to the decision criteria defined in Section 5.2. We follow related publications and execute the MCDA method in three basic steps [203, 54].

- Problem structuring
- Evaluation of options
- Reviewing the decision structure

Multiple MCDA methods have been developed and we use an outranking method. In contrast to approaches such as the analytical hierarchy process (AHP), the Preference Ranking Organization METHod for Enrichment Evaluation (PROMETHEE) is based on a pairwise comparison of the performance scores of different alternatives [204, 162]. If the performance of alternatives on criteria is measured on different scales or using different units of measurement (e.g., kg/MWh and €/capita), an assessment is possible without defining uniform units of measurements and scales for all criteria. Additionally, outranking methods elicit preferences during the decision support process. This is especially helpful for decision problems with many relevant criteria, as it is the case in this project. However, a single decision maker cannot be identified, because many stakeholders take part in the related public discussions about the energy transition, none having the authority over all necessary resources to implement the alternatives [61]. Nonetheless, the method can provide additional support in structuring the decision problem.

At first, a more detailed description of PROMETHEE is presented. The method has been designed for the application and evaluation in a single period. While there have been attempts to incorporate future information in the evaluation [202, 205], the method needs to be further extended to incorporate multiple periods. The developed Multi-Period PROMETHEE approach and its application to the project's decision problem are presented in the subsequent subsections.

11.1 PROMETHEE

Developed by Brans [204, 162] in the 1980s, the outranking method PROMETHEE (Preference Ranking Organization METHod for Enrichment Evaluations) is an established decision support method, that has been applied and further refined in many studies [206]. The first step is the calculation of the difference d_i between the performance scores x_{ij} of two alternatives a and b on each criterion i ($d_i = x_{ia} - x_{ib}$). Depending on the criterion, the performance scores can be minimized or maximized. For example, a maximization is assumed for the criterion *percentage of plants utilizing renewable energies*. An increase in the share of electricity production by renewable plants is therefore assumed to be positive. Contrary to this, a minimization of the criterion $CO_2 - Eq$ means that less emissions contributing to the Global Warming Potential are positive. In PROMTHEE, six different preference functions can be used to express the preference structure of a decision maker. For instance, in the V-shape preference function (see Figure 92), any deviation between alternatives d_i smaller or equal to zero results in no preference (i.e., P = 0) regarding this criterion. A deviation greater than the preference threshold of p results in a strict preference (i.e., P = 1) regarding this criterion. For all preference functions, the value of P is greater or equal to 0 and smaller or equal to 1.

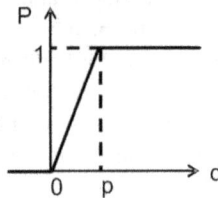

Figure 92: V-shape preference function

Using criteria weights, the resulting preferences are aggregated over all criteria. This yields outranking relations, which show the preference between all pairs of alternatives. These outranking relations are then aggregated over all alternatives, which yields positive outranking flows ϕ^+ and negative outranking flows ϕ^-, representing how much an alternative dominates all the other alternatives, and how much it is dominated by all the other alternatives, respectively. In PROMETHEE I, these positive and negative outranking flows can be used to create a partial ranking of alternatives, where two alternatives can be considered incomparable. Aggregating ϕ^+ and ϕ^- to the net outranking flow ϕ^{net}, also called PROMETHEE II, yields a complete ranking. According to this, an alternative a is preferred to an alternative b if $\phi^{net}(a) > \phi^{net}(b)$.

11.2 Multi-Period PROMETHEE (MP-PROMETHEE)

Up to now, no suitable approach for multi-criteria evaluation of long-term transition paths has been published. Therefore, a multi-period PROMETHEE has been developed for NEDS [201] and is described in the following. The objective of MP-PROMETHEE is the evaluation of transitions paths. A path consists of several alternatives, i.e., different configurations of the power system within the periods (2020, 2030, 2040, and 2050). In the transition paths as defined in Section 8.3, there are three alternatives per period, which are evaluated using the same evaluation criteria and preference thresholds in each of these periods. The performance scores change over time (see Table 35). If changing stakeholder preferences among the criteria need to be modeled, the weights of the criteria can be varied in-between the individual periods.

For the evaluation of the paths and for identifying the best path from today's perspective, the MP-PROMETHEE method is structured into three steps. The first step falls into the *problem structuring* phase of the MCDA process, while the second and third steps are part of the *evaluation of options* phase.

Step 1: Multi-Period problem formulation

In contrast to a single-period problem formulation, the alternatives are defined for several periods. Selected results of the simulation and optimization models described in Section 10 are performance scores for the sustainability criteria. The preference thresholds depend on the analyzed alternatives. In a *single-period evaluation*, following [207], preference and indifference thresholds can be estimated as a percentage of the maximum difference between the performance scores for all alternatives $max\{x_{ij,t}\} - min\{x_{ij,t}\}$ for each criterion i and in each period t. For example, the indifference thresholds q_i can be estimated as $5 - 15\%$ and the preference thresholds p_i as $10 - 30\%$ of this difference. In a *multi-period evaluation*, to account for possible outliers or large deviations of the performance scores in-between the different periods, it is suggested to consider all alternatives in all periods for the calculation of the preference thresholds, using the maximum difference between the performance scores on all alternatives over all periods $max\{max\{x_{ij,t}\} - min\{x_{ij,t}\}\}$.[1] The decision table such as given in Table 35 for Scenario 3 can be compiled using these values. Next to the performance scores for

---

[1] The following simplified decision problem illustrates this necessity: Consider a decision problem where one of the criteria are costs, which are to be minimized. There are two decision periods and the costs are distributed very unevenly. For example, the maximum differences between the performance scores are 10,000€ in the first period and 1,000€ in the second period. Using a single-period evaluation to calculate the p_i, there is no relation between the performance scores of the alternatives of the first period and those of the second period. For the preference threshold, this means that a difference of at least 300€ (following the 30% recommendation by [210]) in the performance scores of two alternatives would be sufficient for strict preference in the second period. In the first period, this difference would have to be 3,000€. Therefore, we suggest a difference of 3,000€ in both the first and the second period as the preference threshold for strict preference.

the individual criteria, Table 35 also includes the thresholds p_i, calculated as 20% of the maximum difference between the performance scores for all alternatives within a period and the weights for each sustainability dimension. The weights are equal for each of the four sustainability dimensions (see Section 0 for a derivation of these values).

Step 2: Assessment of alternatives in each time-step

The alternatives are evaluated separately for each period with PROMETHEE [162, 204], resulting in ϕ^+, ϕ^- and ϕ^{net} for each alternative. For a more detailed evaluation, the single criterion ϕ^+, ϕ^- and ϕ^{net} flows are also calculated. These single criterion flows represent how strongly one alternative dominates another when only one criterion is considered [162]. For example, the ϕ^+ single criterion flow of the criterion *percentage of plants utilizing renewable energies* in the year 2020 for alternative 1 is 0.005. For the criterion *grid efficiency*, the single criterion flow is 0.026. Together, these scores are aggregated to the score attributed to the technical sustainability dimension, which is 0.031 for ϕ^+ in the year 2020 for alternative 1. By comparing this with the total ϕ^+ score of other criteria and alternatives, very detailed assessments can be made.

To provide a more compact overview of the decision problem we aggregate the single criterion results, for the ϕ^+, ϕ^-, ϕ^{net}- flows for each of the four sustainability dimensions (e.g., see Table 34 for the ϕ^+ flows).

Table 34: Result of the ϕ^+ flows for the individual sustainability dimensions for each period and the three paths.

		2020	2030	2040	2050
($a1$)	Technical	0.031	0.009	0.000	0.000
	Social	0.030	0.143	0.023	0.018
	Environmental	0.011	0.179	0.052	0.000
	Economic	0.042	0.000	0.000	0.000
($a2$)	Technical	0.059	0.066	0.188	0.065
	Social	0.024	0.063	0.124	0.188
	Environmental	0.025	0.067	0.107	0.158
	Economic	0.000	0.206	0.166	0.154
($a3$)	Technical	0.000	0.072	0.099	0.098
	Social	0.072	0.066	0.025	0.096
	Environmental	0.048	0.037	0.017	0.104
	Economic	0.084	0.113	0.109	0.103

Step 3: Aggregation of evaluations and sensitivity analysis

In the final step, the evaluations of the alternatives are aggregated along the transition paths Ψ. In a transition graph (see Section 8.3), a path

$\Psi = (a_{j,1}, a_{j,2}, \ldots, a_{j,T})$ is a sequence of alternatives, which are connected with edges over all periods $t \in \{1, \ldots, T\}$. We put up for discussion two aggregation variants, which we illustrate for the ϕ^{net} values. The variants allow investigating the decision problem from two different perspectives. Both variants lead to a normalized evaluation $\phi^{net}(\Psi) \in [-1, 1]$, so that the flows can be interpreted in the same way as the original ϕ^{net} flows of the alternatives: the higher the net flow, the better the evaluation of a path.

Variant I is the arithmetic mean of the ϕ^{net} flows of all alternatives in a path. This implies equal weights of alternatives from different periods. For example, if the objective were to identify the transition toward a sustainable power generation system, it would be reasonable to demand that the assessments of sustainability in the individual years are equally considered.

$$\phi^{net}_{average}(\Psi) = \frac{1}{T} \cdot \sum_{a \in \Psi} \phi^{net}(a) \tag{11-1}$$

Variant II is the average of discounted ϕ^{net} flows, where r is a discount factor. An open question, however, is, what discount factors are suitable. In finance, the Net Present Value (NPV) incorporates the time value of money, which states that earlier cash flows are preferred to later cash flows [208]. For instance, it is used to calculate the net present value of long-term investments such as the construction and operation of wind turbines, with a projected lifetime and investment horizon of up to 20 years. Variant II can be used in a decision problem with a long planning horizon, where the internal uncertainty of the energy system model results is considerably high.

$$\phi^{net}_{averageDis}(\Psi, r) = \frac{1}{T} \cdot \sum_{a \in \Psi} \frac{\phi^{net}(a)}{(1+r)^t} \tag{11-2}$$

The net present value method has also been applied to non-financial values, e.g., to discount future CO_2-emissions [209]. We apply a similar reasoning: Assuming a fixed assessment horizon, positive evaluations (i.e., higher ϕ^{net} values) occurring in earlier periods will prevail for a longer time. This is based on the assumption that a time value is attributed to non-monetary gains as well. Additionally, this takes into account that the performance scores of later periods, and the assessments of them, are more uncertain than those of earlier periods are.

After the aggregation with one of the variants presented above, a sensitivity analysis should be conducted to gain further insight into the decision problem. For variant II, if a value of r cannot be determined, a sensitivity analysis can be conducted to analyze different degrees of uncertainty associated with future assessments.

Table 35: Multi-period decision table for the evaluation of transition paths

Sub-Objective	Criteria	Unit	P-Value	weight	Objective	2020			2030			2040			2050		
						S3 A1	S3 A2	S3 A3	S3 A1	S3 A2	S3 A3	S3 A1	S3 A2	S3 A3	S3 A1	S3 A2	S3 A3
Technical	percentage of plants utilizing renewable energies	%	0.006	0.125	max	0.497	0.497	0.497	0.695	0.723	0.723	0.849	0.857	0.861	0.929	0.942	0.943
	grid efficiency	share of output %	0.0022	0.125	max	0.963	0.964	0.962	0.964	0.964	0.964	0.956	0.967	0.964	0.960	0.960	0.960
	import quota for energy sources used	%	0.006	0.036	min	0.112	0.113	0.122	0.102	0.101	0.130	0.088	0.086	0.092	0.031	0.031	0.039
	ratio of wage to capital income	%	0.0006	0.0357	min	0.992	0.992	0.992	0.989	0.992	0.989	0.989	0.988	0.988	1.002	1.001	1.001
Social	share of expenditure on electricity of total consumption expenditure	%	0.000	0.036	min	0.011	0.011	0.011	0.011	0.010	0.010	0.010	0.010	0.010	0.010	0.010	0.010
	behavioral adaptation costs	€/capita	0.343	0.0357	min	-	-	-	0.159	0.003	0.091	0.869	0.150	0.119	1.822	0.107	0.898
	particulate matter formation	kg PM10-eq/MWh	0.014	0.036	min	0.383	0.384	0.383	0.301	0.372	0.372	0.265	0.266	0.272	0.231	0.214	0.220
	photochemical oxidant formation	kg NMVOC/MWh	0.005	0.036	min	0.444	0.443	0.442	0.431	0.457	0.458	0.324	0.321	0.321	0.283	0.272	0.274
	human toxicity	kg 1,4-DCB-eq/MWh	9.980	0.036	min	488.725	488.172	486.802	387.626	433.396	437.523	120.033	109.196	110.304	93.434	82.031	87.762
Environmental	metal depletion	kg Fe-eq/MWh	0.273	0.036	min	13.270	13.139	13.143	17.170	15.948	16.469	18.826	17.923	18.282	19.472	18.104	19.010
	fossil depletion	kg oil-eq/MWh	0.526	0.036	min	90.004	89.644	89.551	88.193	86.678	86.846	45.100	43.369	42.719	29.633	27.114	27.005
	climate change	kg CO2-eq/MWh	3.267	0.036	min	355.249	354.606	353.838	315.344	331.002	331.680	133.620	128.652	126.787	90.876	84.796	84.556
	terrestrial acidification	kg SO2-eq/MWh	0.093	0.036	min	1.741	1.753	1.747	1.130	1.597	1.591	1.029	1.044	1.082	0.785	0.671	0.706
	freshwater eutrophication	kg P-eq/MWh	0.018	0.036	min	0.772	0.772	0.770	0.589	0.673	0.676	0.118	0.106	0.105	0.070	0.062	0.065
	terrestrial ecotoxicity	kg 1,4-DCB-eq/MWh	0.003	0.036	min	0.080	0.080	0.080	0.079	0.096	0.096	0.097	0.098	0.100	0.097	0.095	0.096
	agricultural land occupation	m²/MWh	0.328	0.036	min	7.762	7.816	7.787	6.003	7.641	7.630	6.322	6.414	6.562	5.457	4.963	5.137
Economic	real gross domestic product	1,000 €/capita	0.012	0.125	max	34.977	34.976	34.977	37.265	37.284	37.275	39.774	39.812	39.793	42.508	42.566	42.537
	costs of electricity production and grid expansion	€/MWh	2.7185	0.125	min	78.694	80.259	77.686	95.953	82.361	83.728	76.830	75.045	73.002	69.377	68.104	67.866

11.3 Results

In the following section, MP-PROMETHEE is exemplarily applied to evaluate the sustainability of the transition paths in scenario 3, which were defined in Section 8.3. The decision table (Table 34) comprises decision relevant results from modeling and simulation.

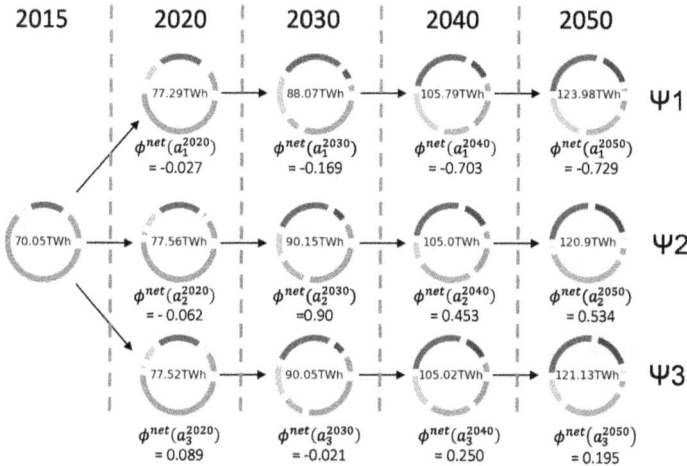

| 2015 | 2020 | 2030 | 2040 | 2050 |

77.29TWh → 88.07TWh → 105.79TWh → 123.98TWh $\Psi 1$

$\phi^{net}(a_1^{2020})$ = -0.027 $\phi^{net}(a_1^{2030})$ = -0.169 $\phi^{net}(a_1^{2040})$ = -0.703 $\phi^{net}(a_1^{2050})$ = -0.729

70.05TWh → 77.56TWh → 90.15TWh → 105.0TWh → 120.9TWh $\Psi 2$

$\phi^{net}(a_2^{2020})$ = - 0.062 $\phi^{net}(a_2^{2030})$ =0.90 $\phi^{net}(a_2^{2040})$ = 0.453 $\phi^{net}(a_2^{2050})$ = 0.534

77.52TWh → 90.05TWh → 105.02TWh → 121.13TWh $\Psi 3$

$\phi^{net}(a_3^{2020})$ = 0.089 $\phi^{net}(a_3^{2030})$ = -0.021 $\phi^{net}(a_3^{2040})$ = 0.250 $\phi^{net}(a_3^{2050})$ = 0.195

Figure 93: PROMETHEE II results for the individual periods for the three alternatives in scenario 3. For reference, the pie chart depicts the annual energy production and the energy mix (gray: fossil fuel, green: biogas, dark blue: wind offshore light blue: wind onshore, yellow: PV-rooftop, orange: ground-mounted PV)

Figure 93 shows the PROMETHEE II results for the four evaluation periods and the alternatives:

- A1: Local power generation with flexible energy demand; focus on onshore wind power and rooftop PVs ("decentral")
- A2: Large-scale storage and power generation; focus on offshore wind power and ground-mounted PVs ("central")
- A3: Middle ground: a mix of all generation and storage technologies ("medium")

In the first period (2020, see Figure 94), A3 has the highest ϕ^{net} flow. A detailed analysis (with stacked bar charts, following the method in [210, 211]) shows that A1 outperforms the other two alternatives on each sustainability dimension but the technical one, where it is outperformed by both A3 and A2. A1 shows the second-best performance. It outperforms A2 in the technical, social and especially in the economic dimensions.

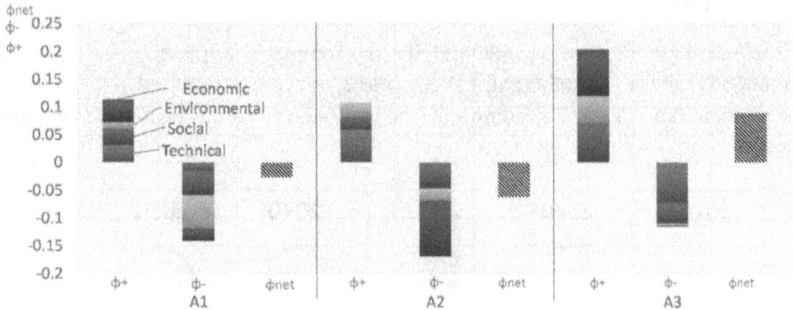

Figure 94: PROMETHEE I & II Results for 2020 split according to sustainability dimensions

In the second period (2030, see Figure 95), A2 outperforms A3 and A1 and is the only alternative with a positive ϕ^{net} flow. A2 and A3 outperform A1 on the criterion *costs of electricity production and grid expansion* in this period, which has a positive effect on the ϕ^{net} flow. While A1 is outperformed on the social and environmental dimension, it still slightly outperforms the other two alternatives on the economic and technical dimensions.

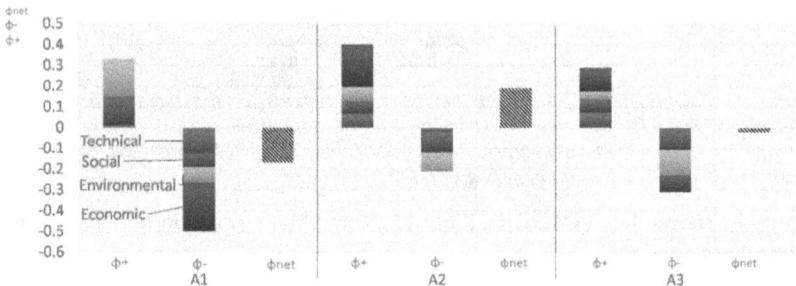

Figure 95: PROMETHEE I & II Results for 2030 split according to sustainability dimensions

In the third period (2040, see Figure 96), the performance of A2 and A3 is very similar. A2 outperforms A3 and both outperform A1 on every sustainability dimension. A1 does no longer outperform the other two alternatives on the social and environmental dimension and therefore its overall performance is reduced.

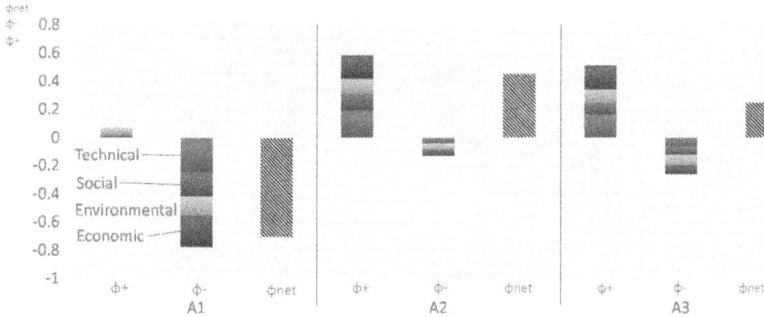

Figure 96: PROMETHEE I & II Results for 2040 split according to sustainability dimensions

In the last period (2050, see Figure 97), the result is similar to that of the previous period (2040): A2>A3>A1 on all sustainability dimensions, except the technical dimension, where A3>A2>A1. Under consideration of all sustainability dimensions and the criteria weights, alternative A2 outperforms both other alternatives.

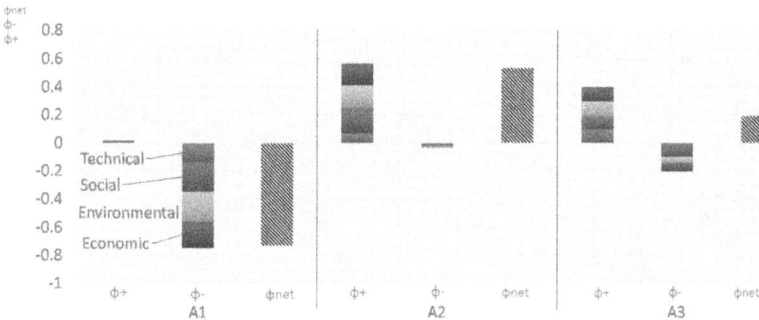

Figure 97: PROMETHEE I & II Results for 2050 split according to sustainability dimensions

Figure 98 shows the aggregated results for the transition paths using aggregation variant I (arithmetic mean). When applying this aggregation method, Ψ2 outperforms the other two paths in every sustainability dimension. At the same time, Ψ3 outperforms Ψ1 in every dimension. When applying this aggregation method, a sensitivity analysis of the criteria weights consequently does not cause a change in the ranking among the three alternatives.

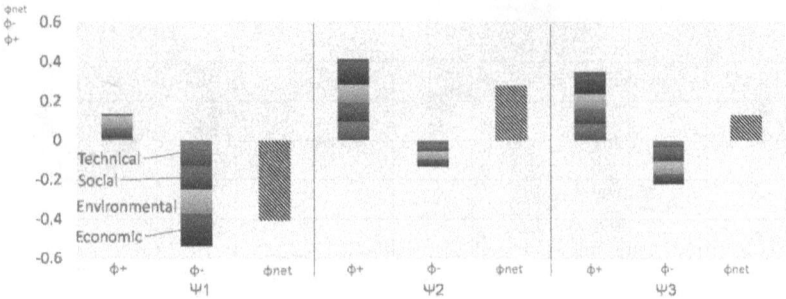

Figure 98: PROMETHEE I & II Results using the aggregation variant I (arithmetic mean)

In Figure 99 an illustration for both presented variants is shown. The figure presents an overview of the assessment of the alternatives in each of the periods as well as the full path evaluation. This figure also offers a more compact view of the performance of the alternatives throughout their paths.

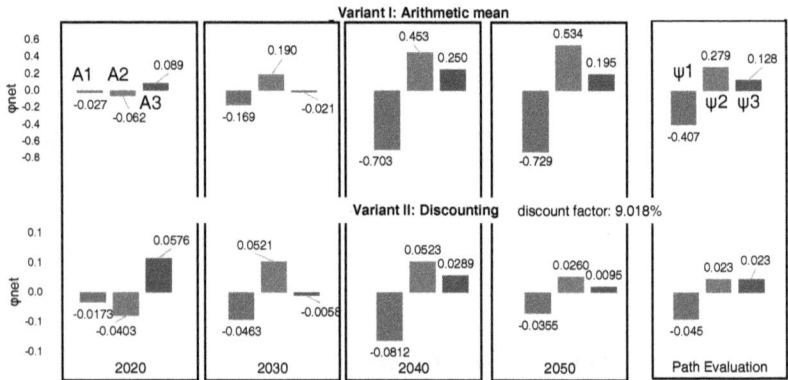

Figure 99: Summarized illustration of the individual evaluations for the alternatives (A1-A3) and paths (Ψ1 – Ψ3)

A time-based sensitivity analysis is possible when applying variant II, the weighted average discount method, where the discount factor r is varied. Figure 100 shows the ϕ^{net} values for an increasing discount factor. The rank of Ψ1 remains unchanged for different values of r. For the other Ψ2 and Ψ3 however, an increased value of r impacts their ranking. The intersection point between Ψ2 and Ψ3 is at 9.018%. If the value of r is further increased, Ψ3 becomes more favorable than Ψ2.

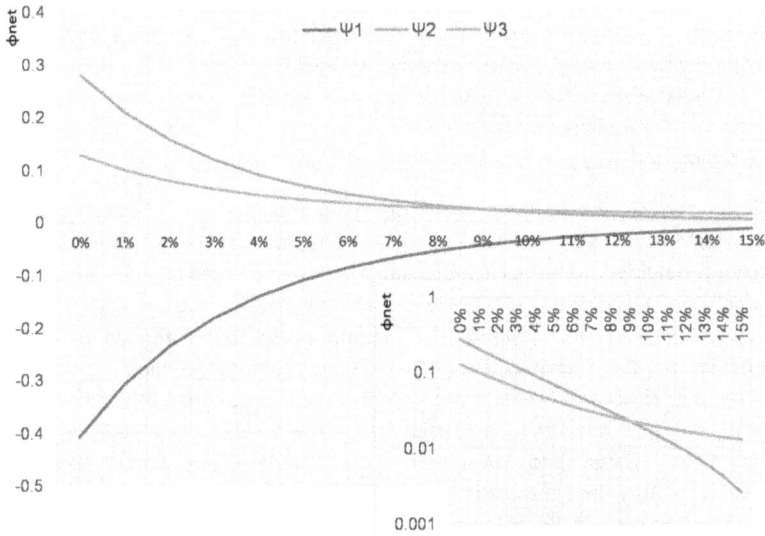

Figure 100: Variation of the discount factor r with an enlarged view (lower right corner) of the intersection point at r = 9,018%

11.4 Discussion

The application of MP-PROMETHEE for an exemplary scenario illustrates how MCDA can be used to integrate the results of energy system models. The advantage of the application of the multi-criteria evaluation method is the formulation of the decision problem in a way that enhances the knowledge about the decision problem. The method achieves this by reducing the problem to a set of alternatives and criteria. The stacked bar charts show the contribution of the sustainability dimensions to the ϕ^+ and ϕ^- flows and allow for a more condensed view of the decision problem.

The example also shows that breaking down the decision problem into multiple periods can have an impact on the evaluation and can uncover that certain alternatives perform better in early periods while being worse of in later periods, or vice versa. This aggregation variant also allows considering the quality and likelihood of the information by quantifying the confidence associated with the assumed developments of the future. Regarding temporal preferences, there are several open questions, which should be put up in the political debate: What is the temporal preference, concerning the depreciation of future performance scores, i.e., are earlier emission reductions at higher costs preferred to later emission reductions at lower costs? Who are the relevant stakeholders?

Concerning the application of MCDA methods to the transition of Lower Saxony's power supply system, the results show that the energy mix is a critical factor for the evaluation of the energy system. For instance, those system configurations that require the largest number of fossil fuel plants as backup always come off worst. In this decision problem, other considered factors such as the length of the transmission grid only played a minor role in the evaluation.

While, in general, the results obtained from the integration of scenario planning, energy system analysis, and MCDA can provide the basis for these kinds of decisions, the results obtained from the exemplary application presented above were obtained for only one scenario and three selected alternatives. For a more comprehensive evaluation and decision support, the method needs to be applied to evaluate alternatives in more scenarios, e.g., the other four scenarios defined in Section 6.2. Furthermore, in accordance with the system boundaries defined in Section 3.2, the different energy system models in this study only represent selected parts of the energy supply system and, consequently, the criteria hierarchy and performance scores used for the evaluation can only support decisions concerning the corresponding parts of the system.

Decisions in energy policy and the energy sector can be supported with MP-PROMETHEE. The long-term view brings important criteria into focus that otherwise would have been overlooked. The results can be used to question why some alternatives come off better than others do. If decentralized solutions are desired in the political debate, one has to consider whether and how one can mitigate their weaknesses or further develop their strengths in comparison to other alternatives.

12. Conclusion

The main result of the project NEDS is the Process for Integrated Development and Evaluation of Energy Scenarios (PDES). This process aims to support decisions in energy policy and the energy sector on the way to fulfilling the climate change goals set by the Paris Agreement and the targets of the Energy Sources Act (EEG). These goals demand increasing amounts of electricity coming from renewable energy sources, which leads to many changes in the energy system affecting the economy, environment, policy, technology, and society. Thus, interdisciplinarity plays an important role in the PDES.

The PDES relies on scenario planning, energy system analysis, and multi-criteria analysis. Scenario planning provides a structured process for thinking about possible future developments and developing qualitative future scenarios. A fundamental step for the integration of the three methods in the PDES is the quantification of the assumptions made in the future scenarios to enable a quantitative examination with energy system analysis, which allows looking in more detail into the functionality and characteristics of the energy system. To take account of the interdisciplinarity needed in the PDES, simulation models from different disciplines are coupled to examine the energy system. Multi-criteria decision analysis (MCDA) is used to help structure the decision problem, evaluate the results, and support decision-making.

The PDES is subdivided into four subsequent phases (see Section 4.1): (1) Preparatory Steps; (2) From Story to Simulation; (3) Modeling and Simulation; (4) Sustainability Evaluation. In *(1) Preparatory Steps*, the concept of sustainability as well as concrete, measurable sustainability evaluation criteria (SEC) are defined for a particular problem. In addition, qualitative future scenarios describing future developments are further developed for this problem. In *(2) From Story to Simulation*, the future scenarios are quantified in such a way that quantitative results can be gained for the SEC directly or via subsequent energy system analysis. To that end, quantitative assumptions for alternatives and external scenarios need to be defined for each considered year. These quantified assumptions represent the transition paths. The assumptions are used in *(3) Modeling and Simulation* to obtain performance scores for the transition paths. Finally, in *(4) Sustainability Evaluation*, the performance scores are aggregated in a multi-criteria evaluation, which allows for an aggregated evaluation of transition paths.

For the application of the PDES, an information model has been developed, which allows the modeling of evaluation goals, data flows, and dependencies in the PDES (see Section 4.2). Future scenarios, quantification, simulation, and evaluation are integrated into the information model. The information model is integrated with Semantic Web technologies to support the following use cases. Firstly, to make its content available via queries, which allows handling complex scenarios with many dependencies and data flows. Secondly, it can assist in the planning of the simulation in the PDES. Thirdly, based on the information model, a scenario database is generated, which stores the quantified attributes and simulation results and makes them available for other models and the evaluation.

To evaluate future energy scenarios, criteria, according to which they should be evaluated, have to be specified. In the NEDS project, our goal was to develop a sustainable future energy system for Lower Saxony. The chosen sustainability concept builds on four dimensions: environmental, social, economic and technical (see Section 5). SEC can be selected based on existing literature or based on approaches, which allow integrating evaluation patterns of groups of interest. Since enabling public participation within PDES was an important aim, we developed SEC and weighting factors for the sustainability dimensions within the MCDA from a qualitative analysis of materials from a public symposium and follow-up interviews with interested citizens (see Section 5.1). The final selection of evaluation criteria is based on integrating both approaches and is limited by the availability of quantifiable attributes and simulation results (see Section 5.2) for the investigated future energy scenario.

Scenario planning is used in the PDES for the structured development of interdisciplinary future scenarios (see Section 6). In NEDS, the scenario planning process was applied, and the most relevant key factors for the development of Lower Saxony's energy supply system identified. For each of them, future projections were defined. Based on these projections, consistent scenarios were defined and five qualitative future scenarios with a textual storyline were created.

For the analysis of the diffusion and adoption of innovations for the energy transition, in Section 7, five innovations relevant for the energy transition in Lower Saxony were derived from the future scenarios and a literature review. These innovations are heat pumps, photovoltaic systems with storage, electric mobility and charging infrastructure, smart meters and dynamic electricity tariffs. Through the evaluation of current studies and an analysis process, in which further actors were derived from qualitative interviews, it was possible to identify relevant adopters and change agents for these innovations. Depending on the innovation, actors of the different groups like private households or distribution system operators took on different roles. Based on these different roles, the elements of diffusion were examined with the help of 46 guided interviews with (potential) adopters and change agents of the innovations. The interviews were conducted and analyzed with a focus on the individual adoption process of private households and the role of change agents in order to cover different perspectives within the diffusion process. As a result, decisive qualitative findings concerning the diffusion of the selected innovations have been brought together in diffusion studies, which in particular provide information about the diffusion-relevant attributes of innovation, central actors of the social system as well as the past and expected future development of diffusion (see Section 7.4). This empirical foundation complements the simulation with valuable qualitative insights of selected components. The results of the diffusion studies of the innovations electric mobility and charging infrastructure, photovoltaics with storage, heat pumps and smart meters were used to shape the transition paths of these innovations in the simulation (see Section 8.3).

The PDES includes a process to quantify the qualitative future scenarios (see Section 8). During this process, attributes, which describe the system under investigation, are defined and classified in endogenous or exogenous. Based on the projections of the future scenarios, the attributes are substantiated for the transition paths with values based on literature research. The endogenous attributes are used to define alternatives, which represent possible courses of action for the stakeholders. From the five defined future scenarios, one scenario, which would fulfill an 80% greenhouse gas emission reduction in reference to 1990, was chosen to be examined in detail with a simulation of different alternatives.

In the simulation, models from different disciplines are coupled in a modular simulation framework (see Section 9). Two of the models are closely coupled in the co-simulation framework mosaik, while the others use CSV files as an interface. Based on the previously quantified attributes, the simulation models calculate the derived attributes, which are necessary for the evaluation of the future scenarios. With this additional quantitative data, aspects can be integrated into the evaluation, which is not covered by existing studies.

The modular *simulation environment eSE* was developed for the investigation as well as the technical and economic evaluation of building systems in the low-voltage level. This enables devices, control systems and the behavior of users to be represented model-based with the possibility of generating and investigating flexible simulation scenarios. In particular, the environment enables the detailed representation of data- and physics-based models and their coupling via defined exchange parameters. The user behavior simulation in eSE builds on a theoretical and empirical analysis of user behavior, which emphasizes the importance of contextual factors for timely variation of user behavior and the importance of behavioral effort of shifting user behavior for analyzing flexibilities for residential buildings. With regard to Lower Saxony, a typical rural and suburban grid region was modeled and simulated in the form of its element's buildings, residential units, users and appliances in order to examine the technical, economic and social effects of a Lower Saxony sustainable energy transformation. Interventions aiming at changing user behavior to influence electrical loads should consider such influencing factors.

12. Conclusion

The electrification and integration of new technologies, such as heat pumps, charging stations, electric vehicles and storage systems into the existing building structures, will result in a change in the active electrical power at the grid connection point, an increase in the temporal loading of the grid assets and lower plannability in the design of low-voltage grids.

In order to investigate the potential benefits of cooperative load scheduling of smart buildings, a smart grid optimization technology at the level of a low-voltage grid was simulated. A distributed optimization process was implemented using the *multi-agent system ISAAC*. In the optimization process, multiple optimization goals were included, such as reducing the peak load, reducing electricity prices, cost optimization in terms of own consumption or minimization of behavioral adaptation efforts for the consumer. As a data basis, feasible schedules of buildings were used, which have been calculated in the building model. Simulation results show that the flexibilities of smart buildings can be used to pursue different targets. One important result was that, if uncontrolled, the concurrent charging of electrical vehicles produces high load peaks. This can be significantly reduced by a control mechanism, which spreads the charging processes throughout a longer time span. Additionally, it was shown that massive feed-in of photovoltaic systems at the low voltage level may play a prominent role in a future energy system. Battery storages may be applied to reduce this problem. However, it was shown, that if battery storages run in a grid-unaware mode solely based on the optimization of own-consumption, they have not enough flexibility to significantly reduce feed-in peaks of photovoltaic systems.

On the macro-level, a modified *Integrated Grid and Market Model* (based on [166]) and *Grid-Planning Algorithm* are able to emulate future target states and their transition paths for the grids in Lower Saxony. With the Integrated Grid and Market Model, the impact of high integration of distributed generators in the transmission grid was investigated. Generally, the need of storage system capacities increases in order to integrate of the rising number of distributed generators in the grid, shown by the results of the simulation. The variation of the capacities increase of the installed storage systems forms the difference between the three alternatives and affects the criteria used to evaluate the alternatives of the scenario. The alternative with the highest amount of storage system capacity in the target state represents the best result with regard to the grid efficiency criterion. Furthermore, the renewable energies in this alternative have the largest share of gross electricity production comparing with the other alternatives. However, the results of all alternatives are very close to each other and in all alternatives, CO_2 emissions are reduced by at least 80%. The grid-planning algorithm is used to investigate the development of grid expansion in the high voltage (HV) and medium voltage (MV) grids. Generally, the simulation shows a need for grid expansion in the HV and MV grids for all alternatives due to the increase of load and feed-in assumptions. Furthermore, the demand for conventional grid reinforcements can be reduced by the modification of the grid topology. In this regard, the coaction of the topology optimization and the optimization of choices of transmission elements is shown. Furthermore, a comparison of the annual costs of the HV and MV grids for the target state and the transition path of the alternatives is given. The annual costs of the transmission paths show a fluctuating behavior depending on each development of load and feed-in assumptions.

To evaluate macroeconomic effects of energy transition policies, a *computable general equilibrium* (CGE) model was applied on a global scale, with Lower Saxony (LSX) and Northwest Germany (NWG), respectively as focus regions. In a first step, interactions between climate policy, particularly the European Union Emissions Trading System (EU ETS), and trade policy were investigated. In a second step, the development of the energy system and economic performance within the future scenarios (FS) and policy alternatives was investigated. The analysis of policy interactions showed that the reduction of tariffs and trade barriers eases LSX's and ROG's CO_2 emissions reductions. The analysis of the FS shows, that the EU ETS can effectively reduce CO_2 emissions in Europe at relatively low costs for NWG and ROG until the emission reduction approaches 80% in 2050. However, if NWG's energy transition fails, costs for climate policy will increase drastically. Results further suggest that a uniform CO_2 price on all sectors would significantly increase welfare and reduce the cost of CO_2 reduction in NWG and ROG.

The *Life-Cycle Assessment* is based on the results of the grid planning and market model and calculates impact indicators for the identified SEC. The report outlined how the data of energy system models can be used to quantify these impact indicators. The results of this model largely contribute to the social and environmental SEC. The results of the Life-Cycle Assessment showed that the energy mix plays a decisive role in determining the emissions of the energy system. In contrast, the transmission grid has a minor impact on most impact indicators. In the exemplary application, the calculated emissions CO_2-Emissions can be reduced by up to 76%. Almost all other indicators decrease through the conversion to a renewable energy supply. Lastly, the method also shows the tradeoffs between the emissions of renewable electricity generation technologies.

The developed paths were evaluated using a newly developed Multi-criteria Decision Analysis method, *MP-PROMETHEE*, which is a modified PROMETHEE approach. With the exemplary evaluation of three alternatives in one selected scenario, the analysis provides a more detailed view of the alternatives' strengths and weaknesses in the individual periods and makes an aggregated evaluation of transition paths possible. By applying the method to the performance scores obtained from the energy system models, the strengths and weaknesses of different energy system configurations can be quantified in relation to the defined sustainability dimensions.

A few limitations need to be taken into consideration, especially regarding the exemplary application of the methodology, analyzing the energy transition of Lower Saxony. First, the involvement of stakeholders, e.g., the "general public", experts from the energy sector, associations, and energy policy was only possible to a limited extent. The PDES in its construction, however, allows for such an integration. Second, the scope of the scenarios and models was limited to Lower Saxony and interdependencies with other states and countries need to be considered. Third, the consistency of all assumptions was only addressed on the level of the key factors but not for all 231 attributes. This is also important to keep in mind for the interpretation of the project results since it is wrong to assume, that because a development results from a simulation, it is also a possible development [212]. For this to be true, all model and scenario assumptions would have to be consistent in themselves and with current scientific knowledge, which is an unrealistic assumption due to necessary simplifications during model specifications. Fourth, the PDES was only executed exemplary for one selected future scenario. To guard against the "cherry picking fallacy" [213], we want to emphasize, that the results cannot be employed to reason for a certain political action recommendation. Finally, the developed system models and quantifiable SEC only represent selected parts of the energy system. This also means that the system analysis and multi-criteria evaluation only pertain to these parts of the energy system.

Some opportunities are definitely present for future research and discussion. The other four developed scenarios should be simulated and evaluated to achieve a more comprehensive assessment. Using these results, it would be interesting to identify strategies in the alternatives, which would work best under different circumstances (external scenarios). This would help to recommend actions for the development of a sustainable energy system. Having developed a generic process description for the integrated development and evaluation of energy scenarios, with an application with more or other models, more aspects could be examined in detail. For example, environmental models, models for the demand of energy in the industry and service sector, or simulation models for social behavior could broaden the scope. In addition, a new execution of the PDES for the energy transition in Lower Saxony would be beneficial, if new framework conditions arise. For example, new climate politics could change assumptions and necessitate new calculations.

13. Acknowledgments

The team of the Chair of Sustainable Production Management (Carl von Ossietzky University of Oldenburg) would like to thank apl. Prof. Dr. Niko Paech, who was in charge of the subproject from April 2015 to October 2017 and developed it further with great commitment. Furthermore, we would like to thank Markus Glötzel for his work in the project from April 2015 to September 2017, especially for conducting five interviews with adopters from the field of PV systems with storage, incorporated in Section 7.4.1. We would also like to thank Arne Lüerßen, whose results of his bachelor thesis on dynamic electricity tariffs in Lower Saxony (2017) were included in Section 7.4.3. Further thanks go to Mikola Krenzel, whose results of his master thesis on barriers and drivers for the diffusion of distributed solar energy storage systems (2019) were used in Section 7.4.1.

The team of the Division of Research Methods and Biopsychology from the Institute of Psychology from the Technische Universität Braunschweig would like to thank Marian Luckhof and Corinna Dietzel, who supported the empirical studies' implementation and analysis.

The team of the OFFIS Institute for Information Technology thanks Dr. Marita Blank and Prof. Dr.-Ing. Astrid Nieße, who took an active part in subproject 4. Astrid Nieße, in particular, contributed the multi-agent framework ISAAC to the project.

The team of the Institute for Environmental Economics and World Trade at the Leibniz Universität Hannover thanks Dr. Frank Pothen for his excellent modeling and project work. The team also thanks Dr. Dorothee Bühler, Judith Soto and Martin Pospisil for their support when finalizing the project and this report.

The teams of the University Duisburg-Essen, Chair of Business Administration and Production Management and the University of Göttingen, Chair of Production and Logistics, thank Katharina Stahlecker, who helped to conceptually develop the PDES, based on the standard procedures in scenario planning and multi-criteria decision analysis.

14. References

[1] The Federal Government, "German Sustainable Development Strategy", 2018.
     [Online]. Available:
     www.bundesregierung.de/resource/blob/975274/1588964/1b24acbed2b731744c2
     ffa4ca9f3a6fc/2019-03-13-dns-aktualisierung-2018-englisch-data.pdf?download=1.
     [Accessed 14 May 2019].

[2] Deutscher Bundestag, „Gesetz für den Ausbau erneuerbarer Energien (Erneuerbare-
     Energien-Gesetz): EEG 2014.", 2014.

[3] A. Harjanne and J. M. Korhonen, "Abandoning the concept of renewable energy",
     Energy Policy, vol. 127, pp. 330-340, 2019.

[4] Niedersächsisches Ministerium für Umwelt, Energie und Klimaschutz, "Lower Saxony
     Ministry for Environment, Energy, and Climate Protection. Leitbild einer
     nachhaltigen Energie- und Klimaschutzpolitik für Niedersachsen", 2016.

[5] M. Faulstich, H.-P. Beck, C. von Haaren, J. Kuck, M. Rode, J. Ahmels, F. Dossola, J.
     zum Hingst, F. Kaiser, A. Kruse, C. Palmas, G. Römer, I. Ryspaeva, H.-H. Schmidt-
     Kanefendt, W. Siemers, R. Simons, J.-P. Springmann and C. Yilmaz, "Szenarien zur
     Energieversorgung in Niedersachsen im Jahr 2050", Niedersächsisches Ministerium
     für Umwelt Energie und Klimaschutz, Hannover, 2016.

[6] D. Keles, D. Möst and W. Fichtner, "The development of the German energy market
     until 2030 - A critical survey of selected scenarios", Energy Policy, vol. 39, no. 2, pp.
     812-825, 2011.

[7] T. Witt, K. Stahlecker and J. Geldermann, "Morphological Analysis of Energy
     Scenarios", International Journal of Energy Sector Management, vol. 12, no. 4, pp.
     525-546, 2018.

[8] F. Zwicky, "The Morphological Approach to Discovery, Invention, Research and
     Construction", in New Methods of Thought and Procedure - Contributions to the
     Symposium on Methodologies, F. Zwicky and A. G. Wilson, Eds., Berlin, Heidelberg,
     New York, Springer, 1967, pp. 273-297.

[9] A. Brandt, K. Rietzler and S. Harms, "Energieland Niedersachsen - Struktur,
     Entwicklung und Innovation in der niedersächsischen Energiewirtschaft", NORD/LB,
     2010.

14. References

[10] U. Fahl and P. Schaumann, "Energie und Klima als Optimierungsproblem - Forschungsvorhaben im Auftrag des BMBF", Institut für Energiewirtschaft und Rationelle Energieanwendung (IER), Stuttgart, 1996.

[11] M. Kralemann, "BUND-Szenario - Energieversorgung in Niedersachsen im Jahr 2050", Bund für Umwelt und Naturschutz Deutschland (BUND), Landesverband Niedersachsen, Hannover, 2018.

[12] D. Snowden, "Complex Acts of knowing: Paradox and descriptive self-awareness", *Journal of Knowledge Management*, vol. 6, no. 2, pp. 100-111, 2002.

[13] W. Weimer-Jehle, J. Buchgeister, W. Hauser, H. Kosow, T. Naegler, W.-R. Poganietz, T. Pregger, S. Prehofer, A. v. Recklinghausen, J. Schippl and S. Vögele, "Context scenarios and their usage for the construction of socio-technical energy scenarios", *Energy*, no. 111, pp. 956-970, 2016.

[14] A. Grunwald, "Der Lebensweg von Energieszenarien - Umrisse eines Forschungsprogramms", in *Energieszenarien*, C. Dieckhoff, W. Fichtner, A. Grunwald, S. Meyer, M. Nast, L. Nierling, O. Renn, A. Voß and M. Wietschel, Eds., Karlsruhe, KIT Scientific Publishing, 2011, pp. 11-24.

[15] S. Steinhilber, J. Geldermann and M. Wietschel, "Renewables in the EU after 2020: a multi-criteria decision analysis in the context of the policy formation process", *EURO Journal on Decision Processes*, vol. 4, no. 1-2, p. 119–155, 2016.

[16] European Union, *"The Treaty on the Functioning of the European Union"*, Brüssel, 2010.

[17] J. S. Schwarz, T. Witt, A. Nieße, J. Geldermann, S. Lehnhoff and M. Sonnenschein, "Towards an Integrated Development and Sustainability Evaluation of Energy Scenarios Assisted by Automated Information Exchange", in *Smart Cities, Green Technologies, and Intelligent Transport Systems*, B. Donnellan, C. Klein, M. Helfert, O. Gusikhin and A. Pascoal, Eds., Cham, Springer International Publishing, 2019, pp. 3-26.

[18] J. Alcamo, "The SAS Approach: Combining Qualitative and Quantitative Knowledge in Environmental Scenarios", in *Environmental Futures - The Practice of Environmental Scenario Analysis*, 1st ed., J. Alcamo, Ed., Amsterdam, Boston, Elsevier, 2008, pp. 123-150.

[19] J. Gausemeier, A. Fink and O. Schlake, "Scenario management: An approach to develop future potentials", *Technological Forecasting and Social Change*, vol. 59, pp. 111-130, 1998.

14. References

[20]   D. Möst and W. Fichtner, "Einführung zur Energiesystemanalyse", in *Energiesystemanalyse - Tagungsband des Workshops "Energiesystemanalyse" vom 27. November 2008 am KIT Zentrum Energie, Karlsruhe,* D. Möst, W. Fichtner and A. Grunwald, Eds., Karlsruhe, Universitätsverlag Karlsruhe, 2009, p. 11–32.

[21]   S. Greco, M. Ehrgott, Figueira and José, "Multiple criteria decision analysis - State of the art surveys", 2nd ed., New York: Springer, 2016.

[22]   J. S. Schwarz, T. Witt, A. Nieße, J. Geldermann, S. Lehnhoff and M. Sonnenschein, "Towards an Integrated Sustainability Evaluation of Energy Scenarios with Automated Information Exchange", in *Proceedings of the 6th International Conference on Smart Cities and Green ICT Systems - Volume 1: SMARTGREENS,* Porto, ScitePress, 2017, pp. 188-199.

[23]   J. Domingue, D. Fensel and J. A. Hendler, "Handbook of Semantic Web Technologies", Berlin Heidelberg: Springer-Verlag, 2011.

[24]   J. S. Schwarz and S. Lehnhoff, "Ontology-Based Development of Smart Grid Co-Simulation Scenarios", in *Proceedings of the EKAW 2018 Posters and Demonstrations Session,* Nancy, France, 2018.

[25]   J. S. Schwarz and S. Lehnhoff, "Ontological Integration of Semantics and Domain Knowledge in Energy Scenario Co-Simulation", in *Proceedings of the 11th International Joint Conference on Knowledge Discovery, Knowledge Engineering and Knowledge Management - (Volume 2),* Vienna, under review.

[26]   F. Schloegl, S. Rohjans, S. Lehnhoff, J. Velasquez, C. Steinbrink and P. Palensky, "Towards a classification scheme for co-simulation approaches in energy systems", in *Proceedings - 2015 International Symposium on Smart Electric Distribution Systems and Technologies, EDST 2015,* 2015.

[27]   J. S. Schwarz, C. Steinbrink and S. Lehnhoff, "Towards an Assisted Simulation Planning for Co-Simulation of Cyber-Physical Energy Systems", in *7th Workshop on Modeling and Simulation of Cyber-Physical Energy Systems,* Montreal, 2019.

[28]   J. Bastian, C. Clauß, S. Wolf and P. Schneider, "Master for Co-Simulation Using FMI", in *Proceedings of the 8th International Modelica Conference,* 2011.

[29]   M. Krötzsch, D. Vrandecic, M. Völkel, H. Haller and R. Studer, "Semantic wikipedia", in *World Wide Web Conference 2006 Semantic Web Track,* 2007.

[30]   J. Hellbrück and E. Kals, "Werte, Umweltbewusstsein und Nachhaltigkeit", in *Umweltpsychologie,* J. Hellbrück and E. Kals, Eds., Wiesbaden, VS Verlag für Sozialwissenschaften, 2012, pp. 87-98.

[31] L. Thommen, "Nachhaltigkeit in der Antike: Begriffsgeschichtliche Überlegungen zum Umweltverhalten der Griechen und Römer", in *Graduiertenkolleg Interdisziplinäre Umweltgeschichte*, B. Hermann, Ed., 2010, pp. 9-24.

[32] K. F. Wiersum, "200 years of sustainability in forestry: Lessons from history", *Environmental Management,* vol. 19, no. 3, pp. 321-329, 1995.

[33] R. Döring and K. Ott, "Nachhaltigkeitskonzepte", *Zeitschrift für Wirtschafts- und Unternehmensethik,* vol. 2, no. 3, pp. 315-342, 2001.

[34] International Energy Agency, "Glossary", 2019. [Online]. Available: https://www.iea.org/about/glossary/r/. [Accessed 4 May 2019].

[35] A. Regelous and J. Meyn, "Erneuerbare Energien - eine physikalische Betrachtung", in *PhyDid B-Didaktik der Physik-Beiträge zur DPG-Frühjahrstagung,* 2011.

[36] L. Seghezzo, "The five dimensions of sustainability", *Environmental Politics,* vol. 18, no. 4, pp. 539-556, 2009.

[37] Enquete-Kommission Umwelt, "Abschlußbericht der Enquete-Kommission „Schutz des Menschen und der Umwelt - Ziele und Rahmenbedingungen einer nachhaltig zukunftsverträglichen Entwicklung"", 1998.

[38] C. H. Antunes and C. O. Henriques, "Multi-Objective Optimization and Multi-Criteria Analysis Models and Methods for Problems in the Energy Sector", in *Multiple Criteria Decision Analysis: State of the Art Surveys,* S. Greco, M. Ehrgott and J. R. Figueira, Eds., New York, NY, Springer New York, 2016, pp. 1067-1165.

[39] J. Zoellner, P. Schweizer-Ries and I. Rau, "Akzeptanz Erneuerbarer Energien", in *20 Jahre Recht der Erneuerbaren Energien,* T. Müller, Ed., Baden-Baden, Nomos, 2012, pp. 91 - 107.

[40] P. Mayring, "Qualitative Inhaltsanalyse: Grundlagen und Techniken", 11th ed., Weinheim: Beltz, 2010.

[41] R. L. Brennan and D. J. Prediger, "Coefficient Kappa: Some Uses, Misuses, and Alternatives", *Educational and Psychological Measurement,* vol. 41, no. 3, pp. 687-699, 1981.

[42] D. Gallego Carrera and A. Mack, "Sustainability assessment of energy technologies via social indicators: Results of a survey among European energy experts", *Energy Policy,* vol. 38, no. 2, pp. 1030-1039, 2010.

[43] Z. Hull, "Sustainable development: premises, understanding and prospects", *Sustainable Development,* vol. 16, no. 2, pp. 73-80, 2008.

14. References

[44] World Commission on Environment and Development (WCED), "Report of the World Commission on Environment and Development: Our Common Future", 1987.

[45] M. Hunecke, "Möglichkeiten und Chancen der Veränderung von Einstellungen und Verhaltensmustern in Richtung einer Nachhaltigen Entwicklung", in *Nachhaltigkeit als radikaler Wandel*, H. Lange, Ed., Wiesbaden, VS Verlag für Sozialwissenschaften, 2008, pp. 95-121.

[46] W. Schenler, S. Hirschberg, P. Burgherr, M. Makowski and J. Granat, "Final report on sustainability assessment of advanced electricity supply options", 2009.

[47] V. Teichert, H. Diefenbacher, D. Dümig and S. Wilhelmy, "Indikatoren zur Lokalen Agenda 21", Wiesbaden: VS Verlag für Sozialwissenschaften, 2002.

[48] W. Klöpffer and B. Grahl, "Ökobilanz (LCA)", Weinheim: Wiley-VCH Verlag, 2009.

[49] C. R. Carter and D. S. Rogers, "A framework of sustainable supply chain management: moving toward new theory", *International Journal of Physical Distribution & Logistics Management,* vol. 38, no. 5, pp. 360-387, 2008.

[50] B. Hopwood, M. Mellor and G. O'Brien, "Sustainable development: mapping different approaches", *Sustainable Development,* vol. 13, no. 1, pp. 38-52, 2005.

[51] B. Moldan, S. Janoušková and T. Hák, "How to understand and measure environmental sustainability: Indicators and targets", *Ecological Indicators,* vol. 17, pp. 4-13, 2012.

[52] R. Kothari, "Environment, Technology, and Ethics", in *Ethics of environment and development: Global Challenge, International Response*, J. Engel and J. Engel, Eds., London, Belhaven Press, 1990, pp. 228-237.

[53] T. W. Luke, "Neither sustainable nor development: reconsidering sustainability in development", *Sustainable Development,* vol. 13, no. 4, pp. 228-238, 2005.

[54] V. Belton and T. Stewart, "Multiple Criteria Decision Analysis - An integrated approach", Boston: Kluwer Academic Publishers, 2002.

[55] J.-J. Wang, Y.-Y. Jing, C.-F. Zhang and J.-H. Zhao, "Review on multi-criteria decision analysis aid in sustainable energy decision-making", *Renewable and Sustainable Energy Reviews,* vol. 13, no. 9, pp. 2263-2278, 2009.

[56] J. Gausemeier and C. Plass, "Zukunftsorientierte Unternehmensgestaltung", München: Carl Hanser Verlag, 2014, p. 446.

[57] K. van der Heijden, "Scenarios: The Art of Strategic Conversation", Wiley, 2005.

14. References

[58]   P. Schoemaker, "Scenario Planning: A Tool for Strategic Thinking", *Sloan management review,* vol. 36, pp. 25-40, 1995.

[59]   L. E. Schlange, "Linking futures research methodologies: An application of systems thinking and metagame analysis to nuclear energy policy issues", *Futures,* vol. 27, no. 8, pp. 823-838, 1995.

[60]   H.-J. Appelrath, H. Kagermann and C. Mayer, "Future Energy Grid - Migrationspfade ins Internet der Energie", München: acatech, 2012.

[61]   T. Witt, M. Dumeier and J. Geldermann, "Combining scenario planning, energy system analysis, and multi-criteria analysis to develop and evaluate energy scenarios", *Journal of Cleaner Production,* under review.

[62]   M. Blank, C. Blaufuß, M. Glötzel, J. Minnemann, M. Nebel-Wenner, A. Nieße, F. Pothen, C. Reinhold, J. S. Schwarz, K. Stahlecker, F. Wille, T. Witt, F. Eggert, B. Engel, J. Geldermann, L. Hofmann, M. Hübler, S. Lehnhoff, N. Paech and M. Sonnenschein, "Whitepaper: NEDS Szenarien - Zukunftsszenarien für eine nachhaltige Energieversorgung in Niedersachsen für das Jahr 2050", Juni 2019. [Online]. Available: https://www.neds-niedersachsen.de/uploads/tx_tkpublikationen/Whitepaper-Szenarien-V1.pdf. [Accessed 30 06 2019].

[63]   C. Busse, A. Regelmann, H. Chithambaram und S. Wagner, „Managerial Perceptions of Energy in Logistics: An Integration of the Theory of Planned Behavior and Stakeholder Theory", *International Journal of Physical Distribution & Logistics 47 (6),* pp. 447-471, 2017.

[64]   K. Fichter and J. Clausen, "Erfolg und Scheitern "grüner" Innovationen. Warum einige Nachhaltigkeitsinnovationen am Markt erfolgreich sind und andere nicht", Marburg: Metropolis-Verlag, 2013.

[65]   E. M. Rogers, "Diffusion of innovations", New York, NY: Free Press, 2003.

[66]   M. Schleper and C. Busse, "Toward a standardized supplier code of ethics: development of a design concept based on diffusion of innovation theory", *Logistics Research,* vol. 6, no. 4, pp. 187-216, 2013.

[67]   J. Hauschildt, "Innovationsmanagement, 3. Auflage", München: Verlag Franz Vahlen GmbH, 2004.

[68]   C. Kunz and S. Kirrmann, "Die neue Stromwelt. Szenario eines 100% erneuerbaren Stromversorgungssystems", Agentur für Erneuerbare Energien e.V., Berlin, 2015.

14. References

[69]   WBGU, "Welt im Wandel: Gesellschaftsvertrag für eine Große Transformation [Hauptgutachten]", 2011. [Online]. Available: https://issuu.com/wbgu/docs/wbgu_jg2011?e=37591641/69400318. [Accessed 18 June 2019].

[70]   M. Schlesinger, D. Lindenberger, C. Lutz, P. Hofer, A. Kemmler, A. Kirchner, S. Koziel, A. Ley, A. Piégsa, F. Seefeldt, S. Strassburg, K. Weinert, A. Knaut, R. Malischek, S. Nick, T. Panke, S. Paulus, C. Tode, J. Wagner, U. Lehr and P. Ulrich, "Entwicklung der Energiemärkte – Energiereferenzprognose", Projekt Nr. 57/12 des Bundesministeriums für Wirtschaft und Technologie, Berlin, 2014.

[71]   J. Nitsch, T. Pregger, T. Naegler, D. Heide, D. Luca de Tena, F. Trieb, Y. Scholz, K. Nienhaus, N. Gerhardt, M. Sterner, T. Trost, A. von Oehsen, R. Schwinn, C. Pape, H. Hahn, M. Wickert and B. Wenzel, "Langfristszenarien und Strategien für den Ausbau der erneuerbaren Energien in Deutschland bei Berücksichtigung der Entwicklung in Europa und global", 2012. [Online]. Available: https://www.dlr.de/dlr/Portaldata/1/Resources/bilder/portal/portal_2012_1/leitstud ie2011_bf.pdf. [Accessed 18 June 2019].

[72]   T. Klaus, C. Vollmer, K. Werner, H. Lehmann and K. Müschen, "Energieziel 2050: 100% Strom aus erneuerbaren Quellen", Umweltbundesamt, Dessau-Roßlau, 2010.

[73]   M. Voigtländer, R. M. Henger, H. Haas, M. Schier and T. Just, "Wirtschaftsfaktor Immobilien 2013. Gesamtwirtschaftliche Bedeutung der Immobilienwirtschaft", Gesellschaft für Immobilienwirtschaftliche Forschung, Wiesbaden, 2013.

[74]   M. Ketokivi and T. Choi, "Renaissance of case research as a scientific method", *Journal of Operations Management,* vol. 32, no. 5, pp. 232-240, 2014.

[75]   A. Przyborski and M. Wohlrab-Sahr, "Qualitative Sozialforschung: Ein Arbeitsbuch", 3rd ed., München: Oldenbourg Verlag, 2010.

[76]   J. Gläser and G. Laudel, "Experteninterviews und qualitative Inhaltsanalyse als Instrumente rekonstruierender Untersuchungen", 4th ed., Wiesbaden: VS Verlag für Sozialwissenschaften, 2010.

[77]   R. K. Yin, "Case study research and applications: Design and methods", 6th ed., Los Angeles; London; New Dehli; Singapore; Washington DC; Melbourne: SAGE, 2018.

[78]   J. W. Creswell, "Research Design - Qualitative, Quantitative and Mixed Methods Approaches", 3rd ed., Thousand Oaks: SAGE Publications, Inc., 2009.

[79]   K. Eisenhardt, "Building Theories from Case Study Research", *The Academy of Management Review,* vol. 14, no. 4, pp. 532-550, 1989.

[80] Fraunhofer ISE, "Aktuelle Fakten zur Photovoltaik in Deutschland", 2018. [Online]. Available: https://www.ise.fraunhofer.de/de/veroeffentlichungen/studien/aktuelle-fakten-zur-photovoltaik-in-deutschland.html. [Accessed 4 June 2019].

[81] E3/DC GmbH, "Marktanalyse: Absatzpotenzial für stationäre Batteriespeicher im privaten und gewerblichen Einsatz in Deutschland", 2016. [Online]. Available: https://www.bves.de/wp-content/uploads/2016/06/Marktanalyse-E3DC-Speicherabsatzpotenzial_final.pdf. [Accessed 4 June 2019].

[82] J. Weniger, J. Bergner, T. Tjaden and V. Quaschning, "Dezentrale Solarstromspeicher für die Energiewende", Hochschule für Technik und Wirtschaft HTW, Berlin, 2015.

[83] J. Figgener, D. Haberschusz, K.-P. Kairies, O. Wessels, B. Tepe and D. U. Sauer, "Wissenschaftliches Mess- und Evaluierungsprogramm Solarstromspeicher 2.0. Jahresbericht 2018", Institut für Stromrichtertechnik und Elektrische Antriebe RWTH Aachen, 2018.

[84] S. Gährs, K. Mehler, M. Bost and B. Hirschl, "Acceptance of Ancillary Services and Willingness to Invest in PV-storage-systems", *Energy Procedia,* vol. 73, pp. 29-36, 2015.

[85] KfW, "Markt für Solarstromanlagen in Deutschland hat sich durch Marktanreizprogramm der KfW etabliert", 2019. [Online]. Available: https://www.kfw.de/KfW-Konzern/Newsroom/Aktuelles/Pressemitteilungen-Details_509248.html. [Accessed 28 February 2019].

[86] Agentur für erneuerbare Energien, "Föderal Erneuerbar - Bundesländer mit neuer Energie. Niedersachsen (NI)", n.d.. [Online]. Available: https://www.foederal-erneuerbar.de/startseite. [Accessed 4 June 2019].

[87] Prognos, "Eigenversorgung aus Solaranlagen. Das Potenzial für Photovoltaik-Speicher-Systeme in Ein- und Zweifamilienhäusern, Landwirtschaft sowie im Lebensmittelhandel. Analyse im Auftrag von Agora Energiewende", 2016. [Online]. Available: https://www.agora-energiewende.de/fileadmin2/Projekte/2016/Dezentralitaet/Agora_Eigenversorgung_PV_web-02.pdf. [Accessed 4 March 2019].

[88] V. Quaschning, J. Weniger, J. Bergner and T. Tjaden, "Die Bedeutung von dezentralen PV-Systemen für die deutsche Energiewende", 2015. [Online]. Available: https://www.volker-quaschning.de/downloads/Staffelstein-2015-Quaschning.pdf. [Accessed 4 March 2019].

[89] K.-P. Kairies, D. Haberschusz, D. Magnor, M. Leuthold, J. Badeda and D. U. Sauer, "Wissenschaftliches Mess- und Evaluierungsprogramm Solarstromspeicher. Jahresbericht 2015", Institut für Stromrichtertechnik und Elektrische Antriebe der RWTH Aachen, 2015.

[90]   K.-P. Kairies, D. Haberschusz, J. v. Ouwerkerk, J. Strebel, O. Wessels, D. Magnor, J. Badeda and D. U. Sauer, "Wissenschaftliches Mess- und Evaluierungsprogramm Solarstromspeicher. Jahresbericht 2016", Institut für Stromrichtertechnik und Elektrische Antriebe der RWTH Aachen, 2016.

[91]   J. Figgener, D. Haberschusz, K.-P. Kairies, O. Wessels, B. Tepe, M. Ebbert, R. Herzog and D. U. Sauer, "Wissenschaftliches Mess- und Evaluierungsprogramm Solarstromspeicher 2.0. Jahresbericht 2017", Institut für Stromrichtertechnik und Elektrische Antriebe der RWTH Aachen, 2017.

[92]   H. Wirth, "Aktuelle Fakten zur Photovoltaik in Deutschland", Fraunhofer ISE, Freiburg, 2019.

[93]   S. Schnurre, "Variable Tarife aus dem Blickwinkel der Lastverlagerung", *Energiewirtschaftliche Tagesfragen,* vol. 64, no. 6, pp. 53-57, 2014.

[94]   Ernst & Young, "Kosten-Nutzen-Analyse für einen flächendeckenden Einsatz intelligenter Zähler. Im Auftrag des Bundesministeriums für Wirtschaft und Technologie", 2013. [Online]. Available: http://www.bmwi.de/BMWi/Redaktion/PDF/Publikationen/Studien/kosten-nutzen-analyse-fuer-flaechendeckenden-einsatz-intelligenterzaehler,property=pdf,bereich=bmwi2012,sprache=de,rwb=true.pdf. [Accessed 11 October 2016].

[95]   K.-D. Maubach, "Strom 4.0. Innovationen für die deutsche Stromwende", Wiesbaden: Springer Vieweg, 2015.

[96]   Bundesministerium für Wirtschaft und Technologie, "Smart Metering in Deutschland. Auf dem Weg zum maßgeschneiderten Rollout intelligenter Messsysteme", 2013. [Online]. Available: https://www.bmwi.de/Redaktion/DE/Downloads/Monatsbericht/Monatsbericht-Themen/11-2013-smart-metering.html. [Accessed 4 June 2019].

[97]   Verband der Elektrotechnik Elektronik Informationstechnik, "VDE|FNN begrüßt Gesetz zur Digitalisierung der Energiewende", 2016. [Online]. Available: https://www.vde.com/resource/blob/835714/cc8d7b9d48f2ec4dafc30ef9760746fa/pressemitteilung-pdf-data.pdf. [Accessed 4 June 2019].

[98]   co2online, "Jährlicher Stromverbrauch eines 5-Personen-Haushalts in Deutschland nach Gebäudetyp im Jahr 2017 (in Kilowattstunden)", 2017. [Online]. Available: https://www.stromspiegel.de/stromverbrauch-verstehen/stromverbrauch-5-personen-haushalt/. [Accessed 8 May 2019].

[99]   Bundesamt für Sicherheit in der Informationstechnik, "Marktanalyse zur Feststellung der technischen Möglichkeit zum Einbau intelligenter Messsysteme nach § 30 MsbG. Version 1.0. 31.01.2019", 2019. [Online]. Available: https://www.bsi.bund.de/SharedDocs/Downloads/DE/BSI/SmartMeter/Marktanalys en/Marktanalyse_nach_Para_30_MsbG.pdf;jsessionid=91CFE5EB91E3305975329D4 EC1E07B4B.2_cid360?__blob=publicationFile&v=7. [Accessed 1 February 2019].

[100] C. Nabe, C. Beyer, N. Brodersen, H. Schäffler, D. Adam, C. Heinemann, T. Tusch, J. Eder, C. de Wyl, J.-H. vom Wege and S. Mühe, "Einführung von lastvariablen und zeitvariablen Tarifen", 2009. [Online]. Available: https://docplayer.org/6174239-Einfuehrung-von-lastvariablen-und-zeitvariablen-tarifen.html. [Accessed 20 February 2019].

[101] A. Liebe, S. Schmitt and M. Wissner, "Quantitative Auswirkungen variabler Stromtarife auf die Stromkosten von Haushalten", Wissenschaftliches Institut für Infrastruktur und Kommunikationsdienste, Bad Honnef, 2015.

[102] M. A. Piette, D. Watson, N. Motegi, S. Kiliccote and P. Xu, "Automated Critical Peak Pricing Field Tests: Program Description and Results", Berkeley: Lawrence Berkeley National Laboratory, 2006.

[103] T. J. Gerpott and M. Paukert, "Gestaltung von Tarifen für kommunikationsfähige Messsysteme im Verbund mit zeitvariablen Stromtarifen - Eine empirische Analyse von Präferenzen privater Stromkunden in Deutschland", *Zeitschrift für Energiewirtschaft,* vol. 37, pp. 83-105, 2013.

[104] Enercity AG, "Privatkunden", 2019. [Online]. Available: https://www.enercity.de/privatkunden/index.html. [Accessed 19 February 2019].

[105] L. Budde, "Akzeptanz von variablen Stromtarifen. Ergebnisse einer qualitativen Vorstufe und einer bevölkerungsrepräsentativen Umfrage", 2015. [Online]. Available: https://www.vzbv.de/sites/default/files/downloads/Akzeptanz-variable-Stromtarife_Umfrage-Forsa-vzbv-November-2015.pdf. [Accessed 14 February 2019].

[106] E. Dütschke, M. Unterländer and M. Wietschel, "Variable Stromtarife aus Kundensicht – Akzeptanzstudie auf Basis einer Conjoint-Analyse", Frauenhofer-Institut für System- und Innovationsforschung (Frauenhofer ISI), Competence Center Energietechnologien und Energiesysteme, Karlsruhe, 2012.

[107] L. Karg, K. Kleine-Hegemann, M. Wedler and C. Jahn, "E-Energy Abschlussbericht. Ergebnisse und Erkenntnisse aus der Evaluation der sechs Leuchtturmprojekte", B.A.U.M. Consult GmbH, München/Berlin, 2014.

14. References

[108] Bundesnetzagentur für Elektrizität, Gas, Telekommunikation, Post und Eisenbahn und Bundeskartellamt, "Monitoringbericht 2018", 2019. [Online]. Available: https://www.bundesnetzagentur.de/SharedDocs/Downloads/DE/Allgemeines/Bund esnetzagentur/Publikationen/Berichte/2018/Monitoringbericht_Energie2018.pdf?__ blob=publicationFile&v=3. [Accessed 20 February 2019].

[109] Audax Energie GmbH, "Tarif Spotmarktpreis", 2015. [Online]. Available: https://www.audaxenergie.de/?sec=tarif_spotmarktpreis. [Accessed 20 February 2019].

[110] Fraunhofer IWES/IBP , "Wärmewende 2030, Schlüsseltechnologien zur Erreichung der mittel- und langfristigen Klimaschutzziele im Gebäudesektor. Studie im Auftrag von Agora Energiewende", 2017. [Online]. Available: https://www.agora-energiewende.de/fileadmin2/Projekte/2016/Sektoruebergreifende_EW/Waermewen de-2030_WEB.pdf. [Accessed 2019 June 20].

[111] Bundesverband Wärmepumpe e.V., "Sektorkopplung", 2019. [Online]. Available: https://www.waermepumpe.de/politik/sektorkopplung/. [Accessed 12 February 2019].

[112] Bundesverband Wärmepumpe e.V., "Wärmepumpen in Siedlungen und Quartieren", 2019. [Online]. Available: https://www.waermepumpe.de/waermepumpe/siedlungsprojekte-quartiersloesungen/. [Accessed 12 February 2019].

[113] KMTechnik Merschmann, "Wärmepumpen für Unternehmen", 2019. [Online]. Available: https://www.kmtechnik.de/waermepumpe/waermepumpen-fuer-unternehmen/. [Accessed 12 February 2019].

[114] Bundesverband Wärmepumpe e.V., "Wie funktioniert die Wärmepumpe?", 2019. [Online]. Available: https://www.waermepumpe.de/waermepumpe/funktion-waermequellen/. [Accessed 12 February 2019].

[115] Statistisches Bundesamt, "Bauen und Wohnen. Baugenehmigungen / Baufertigstellungen von Wohn- und Nichtwohngebäuden (Neubau) nach Art der Beheizung und Art der verwendeten Heizenergie, Lange Reihen ab 1980. 2017", 2018. [Online]. Available: https://www.destatis.de/DE/Themen/Branchen-Unternehmen/Bauen/Publikationen/Downloads-Bautaetigkeit/baugenehmigungen-heizenergie-pdf-5311001.pdf?__blob=publicationFile&v=5. [Accessed 25 April 2019].

[116] P. Thomes, "Grundlagen. Elektromobilität – Zukunftstechnologie oder Nischenprodukt?", in *Elektromobilität. Grundlagen einer Zukunftstechnologie*, 2nd ed., Wiesbaden, Springer Vieweg, 2018, pp. 3-29.

14. References

[117] ADAC, "Kostenvergleich: Elektroautos oft überraschend günstig", ADAC e.V., 2018. [Online]. Available: https://www.adac.de/rund-ums-fahrzeug/e-mobilitaet/antrieb/elektroauto-kostenvergleich/. [Accessed 7 February 2019].

[118] Kraftfahrtbundesamt, "Bestand an Pkw am 01. Januar 2018 nach ausgewählten Kraftstoffarten", 2018. [Online]. Available: https://www.kba.de/DE/Statistik/Fahrzeuge/Bestand/Umwelt/2018_b_umwelt_dusl.html?nn=663524. [Accessed 8 February 2019].

[119] BDEW, „Über 16.100 öffentliche Ladepunkte in Deutschland", 2018. [Online]. Available: https://www.bdew.de/presse/presseinformationen/ueber-16100-oeffentliche-ladepunkte-deutschland/. [Zugriff am 11 2 2019].

[120] R. J. Fisher, "Social Desirability Bias and the Validity of Indirect Questioning", *Journal of Consumer Research*, vol. 20, pp. 303-315, 1993.

[121] J. Flauger, "Megafusion auf dem Energiemarkt: Eon will Innogy übernehmen", 10 03 2018. [Online]. Available: https://www.handelsblatt.com/unternehmen/energie/milliarden-transaktion-megafusion-auf-dem-energiemarkt-eon-will-innogy-uebernehmen/21056966.html?ticket=ST-6289105-QV30LvU2oZ1zhcDSZaTb-ap5. [Accessed 13 May 2019].

[122] VKU, "Stadtwerke setzen Energiewende um - Pressemitteilung Nr. 30 / 2017", 2017. [Online]. Available: https://www.vku.de/fileadmin/user_upload/Verbandsseite/Presse/Pressemitteilungen/30_2017_Energiepolitik.pdf. [Accessed 13 May 2019].

[123] M. Schönfelder, P. Jochem and W. Fichtner, "Energiesystemmodelle zur Szenarienbildung - Potenziale und Grenzen", in *Energieszenarien - Konstruktion, Bewertung und Wirkung - "Anbieter" und "Nachfrager" im Dialog*, C. Dieckhoff, W. Fichtner, A. Grunwald, S. Meyer, M. Nast, L. Nierling, O. Renn, A. Voß and M. Wietschel, Eds., Karlsruhe, KIT Scientific Publishing, 2011, pp. 25-40.

[124] Statistisches Bundesamt, "Bevölkerung Deutschlands bis 2060 - 12. koordinierte Bevölkerungsvorausberechnung", Wiesbaden, 2009.

[125] Statistisches Bundesamt, "Bevölkerung Deutschlands bis 2060 - 13. koordinierte Bevölkerungsvorausberechnung", Wiesbaden, 2015.

[126] International Energy Agency, "World Energy Outlook 2015", 2015.

[127] European Comission, "EU Reference Scenario 2016. Energy, Transport and GHD Emissions. Trends to 2050", 2016.

14. References

[128] Arbeitskreis Volkwirtschaftliche Gesamtrechnung der Länder, "Bruttoinladsprodukt, Bruttowertschöpfung in den Ländern der Bundesrepublik Deutschland von 2000 bis 2014.", 2015.

[129] M. Fischedick, A. Grunwald, W. Canzler, C. Dieckhoff, G. Hirsch Hadorn, P. Kasten, T. Requate, M. Robinius, D. Thrän, D. Vetter and J.-P. Voß, "Pfadabhängigkeiten in der Energiewende - Das Beispiel Mobilität", München: acatech - Deutsche Akademie der Technikwissenschaften e. V., 2017.

[130] Bundesministerium für Umwelt, Naturschutz und nukleare Sicherheit, "Klimaschutzbericht 2018", 2019.

[131] P. Gerbert, P. Herhold, J. Burchardt, S. Schönberger, F. Rechenmacher, A. Kirchner, A. Kemmler and M. Wünsch, "Klimapfade für Deutschland", 2018.

[132] C. Steinbrink, M. Blank-Babazadeh, A. El-Ama, S. Holly, B. Lüers, M. Nebel-Wenner, R. Ramírez Acosta, T. Raub, J. S. Schwarz, S. Stark, A. Nieße and S. Lehnhoff, "CPES Testing with MOSAIK: Co-Simulation Planning, Execution and Analysis", *Applied Sciences,* vol. 9, no. 5, 2019.

[133] E. R. Frederiks, K. Stenner and E. V. Hobman, "The socio-demographic and psychological predictors of residential energy consumption: A comprehensive review", *Energies,* vol. 8, no. 1, pp. 573-609, 2015.

[134] W. Poortinga, L. Steg and C. Vlek, "Values, environmental concern, and environmental behavior: A study into household energy use", *Environment and Behavior,* vol. 36, no. 1, pp. 70-93, 2004.

[135] B. F. Skinner, "Selection by Consequences", *Science,* vol. 213, no. 4507, pp. 501-504, 1981.

[136] FDZ der Statistischen Ämter des Bundes und der Länder, *"Zeitverwendungserhebung 2012/2013",* 2013.

[137] Statistisches Bundesamt, "Zeitverwendungserhebung ZVE 2012/2013", Wiesbaden, 2016.

[138] F. Wille and F. Eggert, *"Identifying contextual factors influencing behavioural variability of energy related behaviours in households",* manuscript in preparation.

[139] Statistisches Bundesamt, "WIE DIE ZEIT VERGEHT - Analysen zur Zeitverwendung in Deutschland", in *Beiträge zur Ergebniskonferenz der Zeitverwendungserhebung 2012/2013,* Wiesbaden, 2017.

[140] V. I. Levenshtein, *"Binary codes capable of correcting deletions, insertions, and reversals",* vol. 10, 1966, pp. 707-710.

[141] L. Kaufman and P. J. Rousseuw, "Finding Groups in Data: An Introduction to Cluster Analysis", New York: John Wiley & Sons, Inc., 1990.

[142] E. K. Morris, "Some reflections on contexualism, mechanism, and behavior analysis", *The Psychological Record,* vol. 47, pp. 529-542, 1997.

[143] E. K. Morris, "Mechanism and Contextualism in Behavior Analysis: Just Some Observations", *The Behavior Analyst,* vol. 16, no. 2, pp. 255-268, 1993.

[144] P. C. Stern, "Psychology and the science of human-environment interactions", *American Psychologist,* vol. 55, no. 5, pp. 523-530, 2000.

[145] G. Schuitema, L. Ryan and C. Aravena, "The Consumer's Role in Flexible Energy Systems: An Interdisciplinary Approach to Changing Consumers' Behavior", *IEEE Power and Energy Magazine,* vol. 15, no. 1, pp. 53-60, 2017.

[146] C. Eid, E. Koliou, M. Valles, J. Reneses and R. Hakvoort, "Time-based pricing and electricity demand response: Existing barriers and next steps", *Utilities Policy,* vol. 40, pp. 15-25, 2016.

[147] J. Pierce, D. J. Schiano and E. Paulos, "Home, Habit, and Energy: Examining Domestic Interactions and Energy Consumption", in *CHI 2010: Home, Eco, Behavior,* Atlanta, GA, USA, 2010.

[148] D. W. Pierce and C. D. Cheney, "Behavior Analysis and Learning", New York: Routledge, 2017.

[149] J.-N. Audet and L. Lefebvre, "What's flexible in behavioral flexibility?", *Behavioral Ecology,* vol. 28, no. 4, pp. 943-947, 2017.

[150] A. B. Bond, A. C. Kamil and R. P. Balda, "Serial Reversal Learning and the Evolution of Behavioral Flexibility in Three Species of North American Corvids (Gymnorhinus cyanocephalus, Nucifraga columbiana, Aphelocoma californica)", *Journal of Comparative Psychology,* vol. 121, no. 4, pp. 372-379, 2007.

[151] C. Reinhold and B. Engel, "Simulation environment for investigations of energy flows in residential districts and energy management systems", in *Internation ETG Congress,* Bonn, 2017.

[152] D. Fischer, A. Härtl and B. Wille-Haussmann, "Model for electric load profiles with high time resolution for German households", *Energy and Buildings,* no. 92, pp. 170-190, 2015.

[153] C. Reinhold, F. Wille, B. Engel and F. Eggert, "Empirische und Synthetische Lastprognose von nutzerabhängigen Verbrauchsgeräten", in *15. Symposium Energieinnovation,* Graz, 2018.

14. References

[154] M. Maier, "Metaanalyse: Digitalisierung der Energiewende", Agentur für erneuerbare Energien, Forschungsradar Energiewende, 2018.

[155] P. Asmus, "Microgrids, virtual power plants and our distributed energy future", *The Electricity Journal,* vol. 23, no. 10, pp. 72-82, 2010.

[156] V. Coelho, M. Cohen, I. Coelho, N. Liu and F. Guimaraes, "Multi-agent systems applied for energy systems integration: State-of-the-art applications and trends in microgrids", *Applied energy,* vol. 187, pp. 820-832, 2017.

[157] J. Ferber and G. Weiss, "Multi-agent systems: an introduction to distributed artificial intelligence (1)", Addison-Wesley Reading, 1999.

[158] M. Wooldridge and N. Jennings, "Intelligent agents: Theory and practice", *The knowledge engineering review,* vol. 10, no. 2, pp. 115-152, 1995.

[159] A. Nieße and M. Tröschel, "Controlled self-organization in smart grids", *IEEE International Symposium on Systems Engineering (ISSE),* 2016.

[160] C. Hinrichs and M. Sonnenschein, "A distributed combinatorial optimisation heuristic for the scheduling of energy resources represented by self-interested agents", *International Journal of Bio-Inspired Computation,* vol. 8, 2016.

[161] D. Hölker, D. Brettschneider, R. Toenjes and M. Sonnenschein, "Choosing communication technologies for distributed energy management in the smart grid", *2017 IEEE PES Innovative Smart Grid Technologies Conference Europe (ISGT-Europe),* 2017.

[162] J.-P. Brans and Y. d. Smet, "PROMETHEE Methods", in *Multiple criteria decision analysis,* New York, Springer, 2016, pp. 187-219.

[163] A. Kumar, B. Sah, A. R. Singh, Y. Deng, X. He, P. Kumar and R. C. Bansal, "A review of multi criteria decision making (MCDM) towards sustainable renewable energy development", *Renewable and Sustainable Energy Reviews,* vol. 69, pp. 596-609, 2017.

[164] O. Grodzevich and O. Romanko, "Normalization and other topics in multi-objective optimization", in *Proceedings of the Fields–MITACS Industrial Problems Workshop 2006,* 2006.

[165] M. Nebel-Wenner, C. Reinhold, F. Wille, A. Nieße and M. Sonnenschein, "Distributed multi-objective scheduling of power consumption for smart buildings", *Energy Informatics, Springer Open,* 2019, in print.

[166] T. Rendel, C. Rathke, T. Breithaupt and L. Hofmann, "Integrated Grid and Power Market Simulation", in *IEEE PES General Meeting 2012*, San Diego, California, USA, 22.-26. July 2012., 2012.

[167] T. Wolgast, C. Blaufuß and H. L., "Evaluation of a Transition Path to a Future Scenario of Lower Saxony's Energy Supply under Variation of Storage", in *VDE/IEEE Power and Energy Student Summit 2018*, 2018.

[168] M. S. M. Z. J. S. H.H. Thies, „Future structure of rural medium-voltage grids for sustainable energy supply", CIRED Workshop, Paper 128, Lisbon, 2012.

[169] H.-P. B. M. S. Schmiesing J. T. Smolka, „Avoiding MV-network expansion by distributed voltage control", CIRED, 22nd International Conference on Electricity Distribution, Stockholm, 10-13 June 2013.

[170] W. W. A. W. H. Ungrad, „Schutztechnik in Elektroenergiesystemen, Vol2,", Heidelberg: Springer-Verlag Berlin Heidelberg GmbH.

[171] Y. H. F. P. Y. Y. Y. Hongmei, „A combined genetic algorithm/simulated annealing algorithm for large scale system energy integration", *ELSEVIER*, 2000.

[172] G. Schlömer and L. Hofmann, "Optimization of Grid Expansion Measures in Low Voltage Grids using an Integer Linear Optimization Problem", IEEE PES POWERCON 2016, Wollongong, Australia, 2016.

[173] F. Pothen and M. Hübler, "A Forward Calibration Method for New Quantitative Trade Models", *Hannover Economic Papers,* vol. No. 643, 2018.

[174] J. Eaton and S. Kortum, "Technology, geography, and trade", *Econometrica,* vol. 70, no. 5, p. 1741–1779, 2002.

[175] P. Armington, "A theory of demand for products distinguished by place of production.", *IMF StaffPap,* vol. 16, no. 1, p. 159–178, 1969.

[176] F. Pothen and M. Hübler, "The Interaction of Climate and Trade Policy", *European Economic Review,* vol. 107, pp. 1-26, 2018.

[177] L. Caliendo and F. Parro, "Estimates of the trade and welfare effects of NAFTA", *The Review of Economic Studies,* vol. 82, no. 1, pp. 1-44, 2015.

[178] L. Caliendo, E. Rossi-Hansberg, F. Parro and P. Sarte, "The Impact of Regional and Sectoral Productivity Changes on the US Economy", *NBER Working Paper,* vol. 20168, 2014.

[179] F. Alvarez and R. Lucas, "General equilibrium analysis of the Eaton-Kortum model of international trade", *Journal of Monetary Economics,* vol. 54, no. 6, p. 1726–1768, 2007.

[180] L. Bretschger, F. Lechthaler, S. Rausch and L. Zhang, "Knowledge diffusion, endogenous growth, and the costs of global climate policy", *European Economic Review,* vol. 93, p. 47–72, 2017.

[181] E. Van der Werf, "Production functions for climate policy modeling: An empirical analysis", *Energy Economics,* vol. 30, no. 6, p. 2964–2979, 2008.

[182] C. Böhringer and A. Löschel, "Promoting renewable energy in europe: A hybrid computable general equilibrium approach", *Energy Journal,* vol. 27, no. SI, p. 135–150, 2006.

[183] Y. H. H. Chen, S. Paltsev, J. M. Reilly, J. F. Morris and M. H. Babiker, "The MIT EPPA6 Model: Economic Growth, Energy Use, and Food Consumption", *Joint Program on the Science and Policy of Global Change Reports,* vol. 278, 2015.

[184] E. Balistreri, R. Hillberry and T. Rutherford, "Structural estimation and solution of international trade models with heterogeneous firms", *Journal of International Economics,* vol. 83, no. 2, p. 95–108, 2011.

[185] E. Balistreri and R. Hillberry, "Structural estimation and the border puzzle", *Journal of International Economics,* vol. 72, no. 2, p. 451–463, 2007.

[186] C. Anderson and J. Vermunt, "Log-multiplicative association models as latent variable models for nominal and/or ordinal data", *Sociological Methodology,* vol. 30, no. 1, p. 81–121, 2000.

[187] L. Goodman, "Simple models for the analysis of association in cross-classifications having ordered categories", *Journal of the American Statistical Association,* vol. 74, no. 367, p. 537–552, 1979.

[188] E. Commission, „EU ETS Handbook", Brussels, Belgium, 2015.

[189] E. Commission, „National Action Plans, Renewable Energy Directive (2009/28/EC)", Brussels, Belgium, 2009.

[190] W. Klöpffer and B. Grahl, "Ökobilanz (LCA)", Weinheim: Wiley-VCH Verlag, 2009.

[191] ISO, „ISO 14040: Environmental management – Life cycle assessment – Principles and framework", 2006.

[192] S. Hellweg and L. Milà i Canals, "Emerging approaches, challenges and opportunities in life cycle assessment", *Science,* vol. 344, no. 6188, p. 1109–1113, 2014.

[193] ISO, *"ISO 14040: Environmental management – Life cycle assessment – Principles and framework",* 2006.

[194] Ecoinvent, "Ecoinvent version 3.4", Swiss Centre for Life Cycle Inventories (Ecoinvent), St. Gallen, 2018.

[195] A. Arvesen, I. B. Hauan, B. M. Bolsøy and E. G. Hertwich, "Life cycle assessment of transport of electricity via different voltage levels: A case study for Nord-Trøndelag county in Norway", *Applied Energy,* vol. 157, pp. 144-151, 2015.

[196] R. S. Jorge and E. G. Hertwich, "Environmental evaluation of power transmission in Norway", *Applied Energy,* vol. 101, pp. 513-520, 2013.

[197] M. Huijbregts, Z. Steinmann, P. Elshout, G. Stam, F. Verones, M. Viera, A. Hollander, M. Zijp and R. van Zelm, "ReCiPe 2016: A harmonized life cycle impact assessment method at midpoint and endpoint level", National Institute for Public Health and the Environment, Bilthoven, 2016.

[198] S. Humbert, P. Fantke and O. Jolliet, "Particulate Matter Formation", in *Life cycle impact assessment,* Dordrecht, Springer, 2015, pp. 97-113.

[199] J. V. Tarazona and M. J. Ramos-Peralonso, "Ecotoxicology, Terrestrial", in *Encyclopedia of Toxicology,* Burlington, Elsevier Science, 2014, pp. 1489-1496.

[200] E. Commission, „Questions and Answers on the Commission's proposal to revise the EU Emissions Trading System. MEMO/08/35", Brussels, Belgium, 2008.

[201] T. Witt, M. Dumeier and J. Geldermann, "Multi-criteria Evaluation of the Transition of Power Generation Systems", in *Multikriterielle Optimierung und Entscheidungsunterstützung: Tagungsband GOR Entscheidungstheorie und -praxis 2018,* K. Küfer, S. Ruzika and P. Halffmann, Eds., Wiesbaden, Springer Gabler, 2019, pp. 121-141.

[202] A. Frini and S. Benamor, "Making Decisions in a Sustainable Development Context", *Computational Economics,* vol. 52, pp. 341-385, 2017.

[203] S. French and J. Geldermann, "The varied contexts of environmental decision problems and their implications for decision support", *Environmental Science & Policy,* vol. 8, no. 4, pp. 378-391, 2005.

14. References

[204] J. P. Brans and P. Vincke, "Note-A Preference Ranking Organisation Method: The PROMETHEE Method for Multiple Criteria Decision-Making", *Management Science,* vol. 31, no. 6, pp. 647-656, 1985.

[205] J. Oberschmidt, J. Geldermann, J. Ludwig and M. Schmehl, "Modified PROMETHEE approach for assessing energy technologies", *International Journal of Energy Sector Management,* vol. 4, no. 2, pp. 183-212, 2010.

[206] M. Behzadian, R. B. Kazemzadeh, A. Albadvi and M. Aghdasi, "PROMETHEE: A comprehensive literature review on methodologies and applications", *European Journal of Operational Research,* vol. 200, no. 1, pp. 198-215, 2010.

[207] T. Tsoutsos, M. Drandaki, N. Frantzeskaki, E. Iosifidis and I. Kiosses, "Sustainable energy planning by using multi-criteria analysis application in the island of Crete", *Energy Policy,* vol. 37, no. 5, pp. 1587-1600, 2009.

[208] I. Fisher, "The Rate of Interest", New York: The Macmillan Company, 1907.

[209] J. Daystar, R. Venditti and S. S. Kelley, "Dynamic greenhouse gas accounting for cellulosic biofuels - Implications of time based methodology decisions", *The International Journal of Life Cycle Assessment,* vol. 22, no. 5, pp. 812-826, 2017.

[210] V. Bertsch, "Uncertainty handling in multi-attribute decision support for industrial risk management", Karlsruhe: Karlsruhe University Press, 2008.

[211] J. Mustajoki and R. P. Hämäläinen, "Web-Hipre - A Java Applet For AHP And Value Tree Analysis", in *5th International Symposium on the Analytic Hierarchy Process (ISAHP'99),* 1999.

[212] G. Betz, "Fehlschlüsse beim Argumentieren mit Szenarien", in *Die Energiewende und ihre Modelle,* C. Dieckhoff and A. Leuschner, Eds., Bielefeld, transcript Verlag, 2016, pp. 117-136.

[213] S. O. Hansson, "Evaluating the Uncertainties", in *The Argumentative Turn in Policy Analysis. Reasoning about Uncertainty,* S. O. Hansson and G. Hirsch Hadorn, Eds., Cham, Springer, 2016, pp. 79-104.

15. List of Figures

16. List of Tables

17. Acronyms

AEEI rates	Historical autonomous energy efficiency improvements rates
AHP	Analytical Hierarchy Process
ALT	Alternatives
API	Application Programming Interface
BAC	Behavioral Adaptive Cost
CES	Constant elasticity of substitution
CGE	Computable General Equilibrium Model
CO_2	Carbon Dioxide
COHDA	Combinatorial Optimization Heuristic for Distributed Agents
CSV	Comma-Separated Values
DER	Distributed Energy Resources
EK module	Eaton and Kortum module
EU	European Union
EU ETS	European Union's emissions trading system
FMI	Functional Mockup Interface
FS	Future scenarios
ICT	Information and Communication Technology
GDP	Gross Domestic Product
GHG	Greenhouse Gas
KF	Key Factor (in Scenario Planning)
KfW	Kreditanstalt für Wiederaufbau
kW	Kilowatt
kWp	Peak Power in Kilowatts
LCA	Life-Cycle Assessment
LSX	Lower Saxony
MAS	Multi-Agent System
MCDA	Multi-Criteria Decision Analysis
MP-PROMETHEE	Multi-Period PROMETHEE
NEDS	Nachhaltige Energieversorgung Niedersachsen (sustainable energy supply Lower Saxony)
NPV	Net Present Value
NREAP	National Renewable Energy Action Plans

17. Acronyms

NWG	Northwest Germany - including the states Lower Saxony, Hamburg and Bremen
OWL	Web Ontology Language
PDES	Process for integrated Development and Evaluation of Energy Scenarios
PROMETHEE	Preference Ranking Organization METHod for Enrichment Evaluations
PV	Photovoltaic
RDF	Resource Description Framework
ROG_1	All states in Germany except Lower Saxony
ROG_2	All states in Germany except Lower Saxony, Hamburg and Bremen
SAS	Story-and-Simulation
SEC	Sustainability Evaluation Criteria
SMW	Semantic Media Wiki
SPARQL	SPARQL Protocol And RDF Query Language
VPP	Virtual Power Plant
WTO	World Trade Organization

18. Appendix

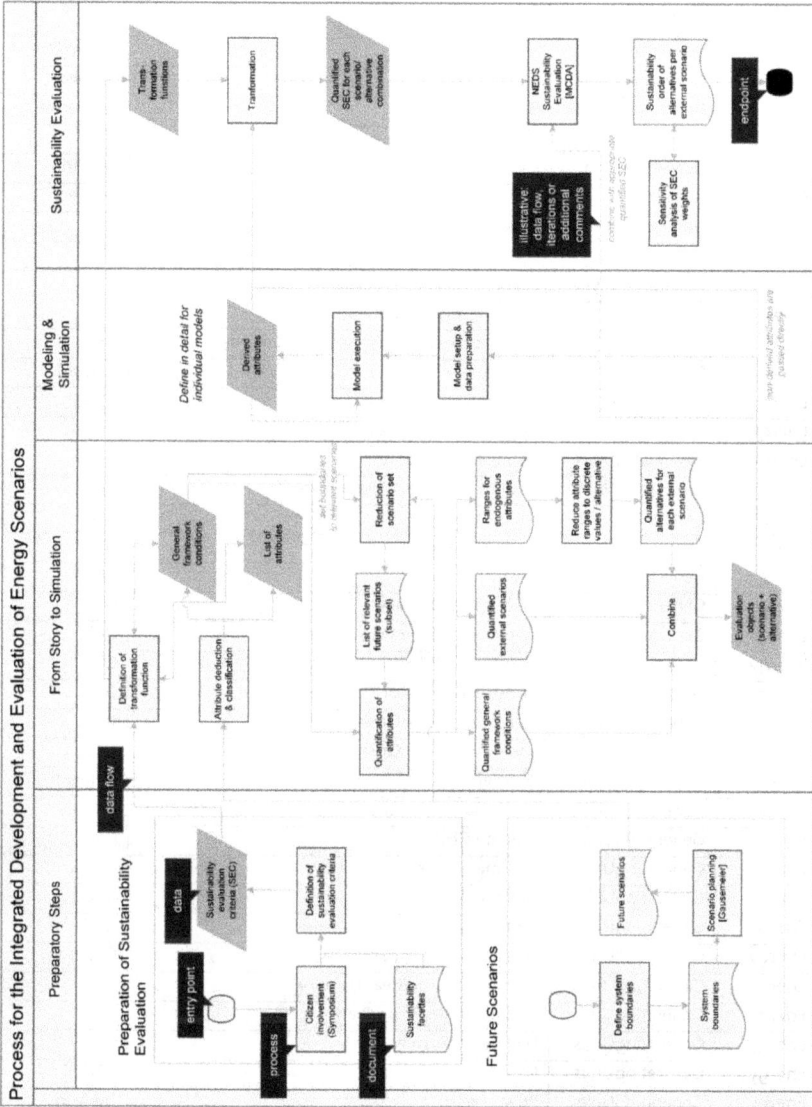

Figure 101: Process diagram of the process for integrated development and evaluation of energy scenarios (PDES) (based on Sustainability Evaluation Process in [17]).

Table 36: List of attributes modeled in PDES (GFC = General Framework Conditions, Scenario = Scenario-Specific)

Domain	Domain Object	Name of attribute	Type
Energy	GeneratingUnits	StartUpCosts	Derived
Energy	GeneratingUnits	FaultRate	Derived
Energy	GeneratingUnits	OperatingCosts	Derived
Energy	GeneratingUnits	GridConnectionPoint	GFC
Energy	GeneratingUnits	ShareOfControllableUnits	Scenario
Energy	GeneratingUnits	PeriodOfUse	Derived
Energy	GeneratingUnits	DistanceToHousingAreas-Biomass	GFC
Energy	GeneratingUnits	DistanceToHousingAreas-Lignite	GFC
Energy	GeneratingUnits	DistanceToHousingAreas-NaturalGas	GFC
Energy	GeneratingUnits	DistanceToHousingAreas-Coal	GFC
Energy	GeneratingUnits	DistanceToHousingAreas-WindTurbines	GFC
Energy	GeneratingUnits	GeneratingCapacity	Derived
Energy	GeneratingUnits	DemandForLand-fossilPlants	GFC
Energy	GeneratingUnits	DemandForLand-GasAndSteamTurbines	GFC
Energy	GeneratingUnits	DemandForLand-PV	GFC
Energy	GeneratingUnits	DemandForLand-WindTurbines	GFC
Energy	GeneratingUnits	Footprint	GFC
Energy	GeneratingUnits	Height-NaturalGas	GFC
Energy	GeneratingUnits	Height-Coal	GFC
Energy	GeneratingUnits	Height-WindTurbines	GFC
Energy	GeneratingUnits	Investment-NaturalGas	GFC
Energy	GeneratingUnits	Investment-PVLarge	GFC
Energy	GeneratingUnits	Investment-PVSmall	GFC
Energy	GeneratingUnits	Investment-WindOffshore	GFC
Energy	GeneratingUnits	Investment-WindOnshore	GFC
Energy	GeneratingUnits	DisasterRisk	Derived
Energy	GeneratingUnits	PowerPlantSchedules	Derived
Energy	GeneratingUnits	Lifetime-CHP	GFC
Energy	GeneratingUnits	Lifetime-Biogas	GFC
Energy	GeneratingUnits	Lifetime-Biomass	GFC
Energy	GeneratingUnits	Lifetime-NaturalGas	GFC
Energy	GeneratingUnits	Lifetime-GeothermalEnergy	GFC
Energy	GeneratingUnits	Lifetime-GasAndSteamTurbines	GFC
Energy	GeneratingUnits	Lifetime-WaterPowerPlant	GFC
Energy	GeneratingUnits	Lifetime-PVSystem	GFC
Energy	GeneratingUnits	Lifetime-WindTurbines	GFC
Energy	GeneratingUnits	PowerFactor	Derived
Energy	GeneratingUnits	AirEmissions-AmmoniacNH3	GFC
Energy	GeneratingUnits	AirEmissions-ArsenicAs	GFC
Energy	GeneratingUnits	AirEmissions-Benzopyrene	GFC

18. Appendix

Energy	GeneratingUnits	AirEmissions-BenzolC6H6	GFC
Energy	GeneratingUnits	AirEmissions-CadmiumCd	GFC
Energy	GeneratingUnits	AirEmissions-HydrogenChlorideHCL	GFC
Energy	GeneratingUnits	AirEmissions-CO2Equivalent	GFC
Energy	GeneratingUnits	AirEmissions-HydrogenFluorideHF	GFC
Energy	GeneratingUnits	AirEmissions-MercuryHg	GFC
Energy	GeneratingUnits	AirEmissions-Iodine129	GFC
Energy	GeneratingUnits	AirEmissions-CarbonMonoxideCO-BioEnergy	GFC
Energy	GeneratingUnits	AirEmissions-CarbonMonoxideCO-WindOffShore	GFC
Energy	GeneratingUnits	AirEmissions-CarbonMonoxideCO-WindOnShore	GFC
Energy	GeneratingUnits	AirEmissions-CarbonDioxideCO2-BioEnergyanlage	GFC
Energy	GeneratingUnits	AirEmissions-CarbonDioxideCO2-WindOffShore	GFC
Energy	GeneratingUnits	AirEmissions-CarbonDioxideCO2-WindOnShore	GFC
Energy	GeneratingUnits	AirEmissions-MethaneCH4-BioEnergy	GFC
Energy	GeneratingUnits	AirEmissions-MethaneCH4-WindOffShore	GFC
Energy	GeneratingUnits	AirEmissions-MethaneCH4-WindOnShore	GFC
Energy	GeneratingUnits	AirEmissions-NitrousOxideN2O	GFC
Energy	GeneratingUnits	AirEmissions-NickelNi	GFC
Energy	GeneratingUnits	AirEmissions-NMVOC-BioEnergy	GFC
Energy	GeneratingUnits	AirEmissions-NMVOC-WindOnShore	GFC
Energy	GeneratingUnits	AirEmissions-PAH	GFC
Energy	GeneratingUnits	AirEmissions-LeadPb	GFC
Energy	GeneratingUnits	AirEmissions-ParticulateMatterPM10	GFC
Energy	GeneratingUnits	AirEmissions-ParticulateMatterPM25	GFC
Energy	GeneratingUnits	AirEmissions-SulphurDioxideSO2	GFC
Energy	GeneratingUnits	AirEmissions-HydrogenSulfideH2S	GFC
Energy	GeneratingUnits	AirEmissions-SulphurDioxideSO2Equivalent	GFC
Energy	GeneratingUnits	AirEmissions-Dust	GFC
Energy	GeneratingUnits	AirEmissions-NitrogenOxidesNOxBioEnergy	GFC
Energy	GeneratingUnits	AirEmissions-NitrogenOxidesNOxWindOffShore	GFC
Energy	GeneratingUnits	AirEmissions-NitrogenOxidesNOxWindOnShore	GFC
Energy	GeneratingUnits	PlanningAreaRequirement-Biomass	GFC
Energy	GeneratingUnits	PlanningAreaRequirement-RooftopPV	GFC
Energy	GeneratingUnits	PlanningAreaRequirement-FossilePlants	GFC
Energy	GeneratingUnits	PlanningAreaRequirement-FreeFieldPV	GFC
Energy	GeneratingUnits	PlanningAreaRequirement-WindTurbines	GFC
Energy	GeneratingUnits	ApparentPower	GFC
Energy	GeneratingUnits	SpecificCO2Emissions-Biomass	GFC
Energy	GeneratingUnits	SpecificCO2Emissions-Lignite	GFC
Energy	GeneratingUnits	SpecificCO2Emissions-GasAndSteamTurbine	GFC
Energy	GeneratingUnits	SpecificCO2Emissions-NuclearEnergy	GFC
Energy	GeneratingUnits	SpecificCO2Emissions-PV	GFC
Energy	GeneratingUnits	SpecificCO2Emissions-Coal	GFC
Energy	GeneratingUnits	SpecificCO2Emissions-WaterPowerPlant	GFC

Energy	GeneratingUnits	SpecificCO2Emissions-WindTurbines	GFC
Energy	GeneratingUnits	Type	GFC
Energy	GeneratingUnits	Efficiency-Biomass	GFC
Energy	GeneratingUnits	Efficiency-Lignite	GFC
Energy	GeneratingUnits	Efficiency-NaturalGas	GFC
Energy	GeneratingUnits	Efficiency-GasAndSteamTurbine	GFC
Energy	GeneratingUnits	Efficiency-NuclearEnergy	GFC
Energy	GeneratingUnits	Efficiency-PV	GFC
Energy	GeneratingUnits	Efficiency-Coal	GFC
Energy	GeneratingUnits	Efficiency-WaterPowerPlant	GFC
Energy	GeneratingUnits	Efficiency-WindTurbines	GFC
Energy	Energymix	Biomass	Endogenous
Energy	Energymix	NaturalGas	Endogenous
Energy	Energymix	WaterPowerPlant	Endogenous
Energy	Energymix	RooftopPV	Endogenous
Energy	Energymix	FreeFieldPV	Endogenous
Energy	Energymix	GeothermalEnergy	Endogenous
Energy	Energymix	WindOffshore	Endogenous
Energy	Energymix	WindOnshore	Endogenous
Energy	Intelligent Equipment	FaultRate	Derived
Energy	Intelligent Equipment	DemandForLand	Derived
Energy	Intelligent Equipment	Investment	Derived
Energy	Intelligent Equipment	Efficiency	Derived
Energy	PowerLines	FailureProbability	Derived
Energy	PowerLines	ExpectedDowntime	Derived
Energy	PowerLines	OperatingCosts	Derived
Energy	PowerLines	Investment	Derived
Energy	PowerLines	Lines-DemandForLand	Derived
Energy	PowerLines	ShortInterruptionFrequency	Derived
Energy	PowerLines	ConductingMaterialAlCu	GFC
Energy	PowerLines	RepairDuration	Derived
Energy	PowerLines	HeightOfPowerLinesMaximumVoltage	GFC
Energy	PowerLines	HeightOfPowerLinesHighVoltage	GFC
Energy	PowerLines	ShareOfCablesMaximumVoltage	Endogenous
Energy	PowerLines	ShareOfCablesHighVoltage	Endogenous
Energy	PowerLines	ShareOfCablesMediumVoltage	Endogenous
Energy	PowerLines	ShareOfCablesLowVoltage	Endogenous
Energy	PowerLines	DepthOfCablesMaximumVoltage	GFC
Energy	PowerLines	DepthOfCablesHighVoltage	GFC
Energy	PowerLines	DepthOfCablesMediumVoltage	GFC
Energy	PowerLines	DepthOfCablesLowVoltage	GFC
Energy	Loads	ShareControllable-Oven	Scenario

Energy	Loads	ShareControllable-Computer	Scenario
Energy	Loads	ShareControllable-ElectricCar	Scenario
Energy	Loads	ShareControllable-Stove	Scenario
Energy	Loads	ShareControllable-TV	Scenario
Energy	Loads	ShareControllable-DishWasher	Scenario
Energy	Loads	ShareControllable-CoffeeMachine	Scenario
Energy	Loads	ShareControllable-FridgeAndFreeezer	Scenario
Energy	Loads	ShareControllable-Storage	Scenario
Energy	Loads	ShareControllable-Dryer	Scenario
Energy	Loads	ShareControllable-Heatpump	Scenario
Energy	Loads	ShareControllable-WashingMachine	Scenario
Energy	Loads	OperatingTime	Derived
Energy	Loads	EnergyEfficiency-Heating-ServiceAndIndustrySector	GFC
Energy	Loads	EnergyEfficiency-Heating-Households	GFC
Energy	Loads	EnergyEfficiency-LightICTAndCooling-ServiceSector	GFC
Energy	Loads	EnergyEfficiency-LightICTAndCooling-Households	GFC
Energy	Loads	EnergyEfficiency-LightICTAndCooling-IndustrySector	GFC
Energy	Loads	EnergyEfficiency-Mobility	GFC
Energy	Loads	EnergyEfficiency-ProcessHeat-Households	GFC
Energy	Loads	EnergyEfficiency-ProcessHeat-ServiceAndIndustrySector	GFC
Energy	Loads	Investment	GFC
Energy	Loads	EfficiencyFactor	Derived
Energy	Loads	GridConnectionPoint	GFC
Energy	Loads	ApparentPower	Derived
Energy	Loads	Type	GFC
Energy	Loads	Power	GFC
Energy	GridExpansion	ExpansionCosts	Derived
Energy	GridExpansion	DemandForLand	Derived
Energy	GridExpansion	GridEfficiency	Derived
Energy	Storage	Aging	GFC
Energy	Storage	ShareControllableStorages-Households	GFC
Energy	Storage	ShareControllableStorages-Mobile	GFC
Energy	Storage	ShareControllableStorages-Grid	GFC
Energy	Storage	CapacityShortTermStorage	Endogenous
Energy	Storage	CapacityLongTermStorage	Endogenous
Energy	Storage	GridConnectionPoint	Derived
Energy	Storage	FailureRate	Derived
Energy	Storage	DemandForLand	Derived
Energy	Storage	StateOfCharge	Derived
Energy	Storage	Investment	Derived
Energy	Storage	EfficiencyFactorCompressedAirStorage	GFC
Energy	Storage	EfficiencyFactorElectrochemicalStorage	GFC

Energy	Storage	EfficiencyFactorPumpStorage	GFC
Energy	Storage	EfficiencyFactorHydrogenStorage	GFC
Energy	SubGrid	BoundsOfVoltageLevel	GFC
Energy	Transformers	FailureRate	Derived
Energy	Transformers	DemandForLand	Derived
Energy	Transformers	Investment	Derived
ICT	Diffusion	NumberOfBuildings	GFC
ICT	Diffusion	NumberOfSmartMeters	Scenario
ICT	Diffusion	NumberOfElectricalConnections	Scenario
ICT	Diffusion	ShareEnergymanagementHouseholds	Scenario
ICT	Diffusion	ShareEnergymanagementIndustry	Scenario
ICT	BEMS	Schedule	Derived
ICT	BEMS	Flexibility	Derived
ICT	BEMS	IndicatorForFossilEnergyUsage	Derived
ICT	ICT-Infrastructure	DistributionStandardsProtocolsAndServices	Endogenous
ICT	SmartMeters	Resolution	GFC
ICT	SmartMeters	Accuracy	GFC
ICT	SmartMeters	Costs	GFC
ICT	SmartMeters	Type	GFC
ICT	SmartMeters	Measurement	GFC
ICT	SmartMeters	DistributionSmartMeter	GFC
ICT	Controller	Schedule	Derived
ICT	Controller	PlanningStrategy	GFC
Market	JobMarket	Salary	Derived
Market	JobMarket	RealIncomeLowerSaxony	Derived
Market	JobMarket	RealIncomeRestOfGermany	Derived
Market	JobMarket	EmploymentInSectors	Derived
Market	CO2-Emissions	Economy	Derived
Market	CO2-Emissions	Industry	Derived
Market	CO2-Emissions	ElectricalPowerPlants	Derived
Market	CO2-Intensity	Economy	Derived
Market	CO2-Intensity	Industry	Derived
Market	CO-Certificates	Price	Derived
Market	Ecology	EnergyFootprint	Derived
Market	Ecology	UsageOfRawMaterials	Derived
Market	Prices	GasPrice	Scenario
Market	Prices	CoalPrice	Scenario
Market	Prices	OilPrice	Scenario
Market	Sectors	ShareOfServiceSectorInGrossValue	Scenario
Market	Sectors	ShareOfAgricultureInGrossValue	Scenario
Market	Sectors	ShareOfManifcaturingIndustryInGrossValue	Scenario
Market	EnergyDemand	EnergyDemand-Households	Derived
Market	EnergyDemand	EnergyDemand-IndustryServiceSectorAndAgriculture	Derived
Market	EnergyPrices	BasicPrice	Derived

Market	EnergyPrices	EEXPrice	Derived
Market	EnergyPrices	CustomerPrice	Derived
Market	EnergyPrices	LevelisedCostsOfElectricity	Derived
Market	Economy	ChangesInEnergyIntensity	Scenario
Market	Economy	GrowthOfEconomy	Scenario
Market	Economy	ImportQuota	Derived
Market	Economy	RealGDP	Derived
User	Behaviour	EquipmentInventory	Scenario
User	Behaviour	OperatingTimeOfEquipment	Scenario
User	Behaviour	FrequencyOfEquipmentUsage	Scenario
User	Behaviour	BehaviouralAdaptationCosts	Derived
User	Sociodemography	AverageSizeOfHousehold	GFC
User	Sociodemography	Inhabitants	GFC
User	Changes	OwnConsumption	Derived
User	Changes	SavingsThroughChangesInBehaviour	Derived
User	Changes	SavingsThroughBetterEfficiency	Derived
User	Changes	ChangesInProfileThroughExternalControl	Derived
User	Changes	ChangesInProfileThroughChangesInBehaviour	Derived
Politics	Subsidies	FeedInRemuneration	GFC

Table 37: Detailed overview of the data used from the Ecoinvent database (version 3.5, 2018); LCIA Method: ReCiPe Midpoint

Power Plant Type	Activity Name (Ecoinvent)	Reference Product
Photovoltaic Rooftop	electricity production, photovoltaic, 3kWp slanted-roof installation, multi-Si, panel, mounted	electricity, low voltage [kWh]
Photovoltaic open ground	electricity production, photovoltaic, 570kWp ground-mounted installation, multi-Si	electricity, low voltage [kWh]
Wind Onshore	electricity production, wind, 1-3MW turbine, onshore	electricity, high voltage [kWh]
Wind Offshore	electricity production, wind, 1-3MW turbine, offshore	electricity, high voltage [kWh]
Biogas	heat and power co generation, biogas, gas engine	electricity, high voltage [kWh]
Hydropower	electricity production, hydro, run-of river	electricity, high voltage [kWh]
Bituminous Coal	heat and power co-generation, lignite	electricity, high voltage [kWh]
Brown Coal	heat and power co-generation, hard coal	electricity, high voltage [kWh]
Nuclear Power	electricity production, nuclear, boiling water reactor	electricity, high voltage [kWh]
Natural Gas	electricity production, natural gas, combined cycle power plant	electricity, high voltage [kWh]
Waste	electricity, from municipal waste incineration to generic market for electricity, medium voltage	electricity, high voltage [kWh]
Transmission grid medium voltage	transmission grid construction, electricity, medium voltage	transmission grid, electricity, medium voltage [km]
Transmission grid high voltage	transmission grid construction, electricity, high voltage	transmission grid, electricity, high voltage [km]